BOB MILLER'S CALC FOR THE CLUELESS

CALC III

OTHER TITLES IN BOB MILLER'S CALC FOR THE CLUELESS

Bob Miller's Calc for the Clueless: Precalc, Second Edition
Bob Miller's Calc for the Clueless: Calc I, Second Edition
Bob Miller's Calc for the Clueless: Calc II, Second Edition

BOB MILLER'S CALC FOR THE CLUELESS

CALC III

Robert Miller

Mathematics Department
City College of New York

McGraw-Hill

New York St. Louis San Francisco Auckland Bogotá
Caracas Lisbon London Madrid Mexico City Milan
Montreal New Delhi San Juan Singapore
Sydney Tokyo Toronto

BOB MILLER'S CALC FOR THE CLUELESS: CALC III

1 2 3 4 5 6 7 8 9 10 11 12 13 14 15 16 17 18 19 20 DOC DOC 9 0 2 1 0 9 8 7

ISBN 0-07-043410-7

Sponsoring Editor: Barbara Gilson
Production Supervisor: Clara Stanley
Editing Supervisor: Maureen Walker
Project Supervision: North Market Street Graphics
Photo: Eric Miller

McGraw-Hill
A Division of The ***McGraw-Hill*** *Companies*

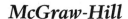

To my wife, Marlene, I dedicate this book and anything else I ever do. I love you. I love you! I LOVE YOU!!!!!

ABOUT BOB MILLER . . . IN HIS OWN WORDS

I received my B.S. and M.S. in math from Brooklyn Poly, now Polytechnic University. After my first class, which I taught as a substitute for a full professor, one student told another upon leaving the room that "at least now we have someone who can teach the stuff." I was forever hooked on teaching. Since then I have taught at Westfield State College, Westfield, Massachusetts; Rutgers; and the City College of New York, where I've been for the last 28½ years. No matter how bad I feel before class, I always feel great after I start teaching. I especially like to teach precalc and calc, and I am always delighted when a student tells me that he or she has always hated math before and could never learn it, but that taking a class with me has made math understandable and even enjoyable. I have a fantastic wife, Marlene; a wonderful daughter, Sheryl; a terrific son, Eric; and a great son-in-law, Glenn. The newest member of our family is my adorable, brilliant granddaughter Kira Lynn, eight days old as of this writing. My hobbies are golf, bowling, bridge, and crossword puzzles. Someday I hope a publisher will allow me to publish the ultimate high school text and the ultimate calculus text so our country can remain number one forever.

To me, teaching math always is a great joy. I hope I can give some of this joy to you.

TO THE STUDENT

This book was written for you: not your teacher, not your next-door neighbor, not for anyone but you. I have tried to make the examples and explanations as clear as I can. However, as much as I hate to admit it, I am not perfect. If you find something that is unclear or should be added to this book, please let me know. If you want a response, or I can help you, your class, or your school, in any precalculus or calculus subject, please let me know, but address your comments c/o McGraw-Hill, Schaum Division, 11 West 19th St., New York City, New York 10011.

If you make a suggestion on how to teach one of these topics better and you are the first and I use it, I will give you credit for it in the next edition.

Please be patient on responses. I am hoping the book is so good that millions of you will write. I will answer.

Now, enjoy the book and learn.

CONTENTS

CHAPTER 1

VECTORS, LINES, PLANES

VECTORS

The first topic we will do is vectors. Instead of first doing everything in two dimensions and then going to three dimensions, we will go to 3D immediately. If you want 2D, you will do ⅓ less arithmetic, while 30D is ten times the arithmetic. Most of the principles are exactly the same.

At the start of most new topics, and vectors are really something new, there is a long introduction. This one is quite tedious. If you don't need it, skip a couple of pages.

Introduction 1: x-y-z in 3D

Points (x,y,z) in 3D require three coordinates. The origin is (0,0,0). In drawing points or 3D curves, positive x is out of the paper, toward you, drawn at approximately a 45° angle; positive y is to the right; and positive z is up. The space is divided into *octants.* The first octant is where x, y, and z are positive—above the x-y plane, to the right of the x-z plane, and above the x-y plane.

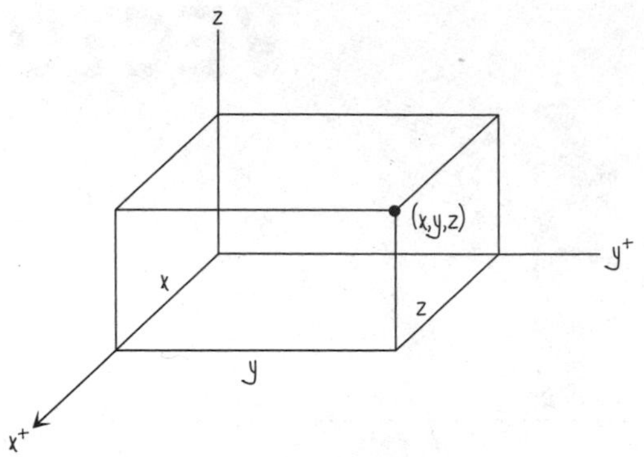

Let us do two points.

A. Point (2,–3,4) is 2 in the positive x direction (toward us), 3 in the negative y direction (left), and 4 in the positive z direction (up).

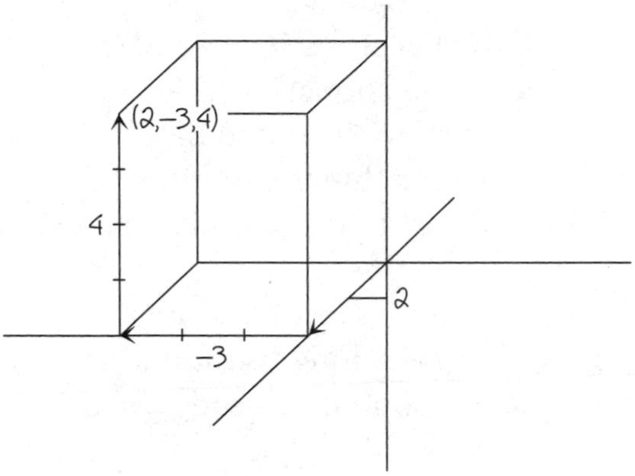

B. Point (–3,4,–5) is 3 in the negative x direction (back 3), 4 in the positive y direction (right 4), and 5 in the negative z direction (down 5).

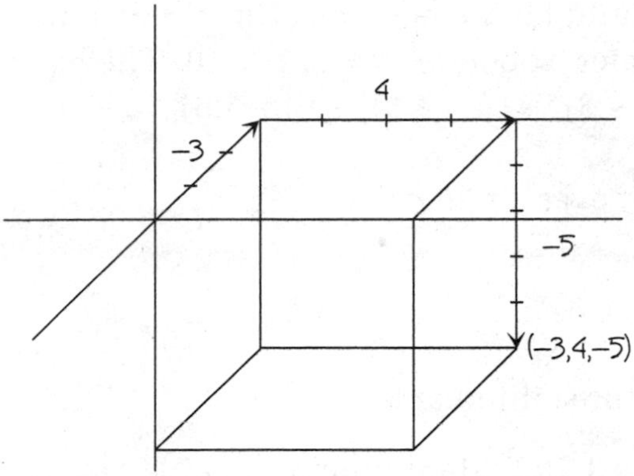

NOTE

Some people have a great deal of difficulty in 3D. Practice will make it go a lot better—so keep trying.

Distance formula (3D extension of Pythagorus)—(x_1,y_1,z_1), (x_2,y_2,z_2); distance

$d = \sqrt{[(x_2 - x_1)^2 + (y_2 - y_1)^2 + (z_2 - z_1)^2]}$

Midpoint formula—$((x_1 + x_2)/2, (y_1 + y_2)/2, (z_1 + z_2)/2)$

Sphere—Center (h,k,l) radius r; $(x - h)^2 + (y - k)^2 + (z - l)^2 = r^2$

Line—A little more complicated—we will do later.

Plane—$Ax + By + Cz = D$, A, B, C all not zero—we will derive later.

EXAMPLE 1—

Given points (1,2,3) and (5,7,−4):

A. Distance $\sqrt{[(5 - 1)^2 + (7 - 2)^2 + (-4 - 3)^2]} = \sqrt{90} = 3\sqrt{10}$

B. Midpoint $((1 + 5)/2, (2 + 7)/2, (-3 - 4)/2) = (3,9/2,-7/2)$

C. Sphere with above points as a diameter

The midpoint would be the center of the circle, and distance is diameter, sooooooo $r = 3/2\sqrt{10}$. The equation would be $(x - 3)^2 + (y - 4.5)^2 + (z + 3.5)^2 = 45/2$ (r squared).

We will sketch several planes.

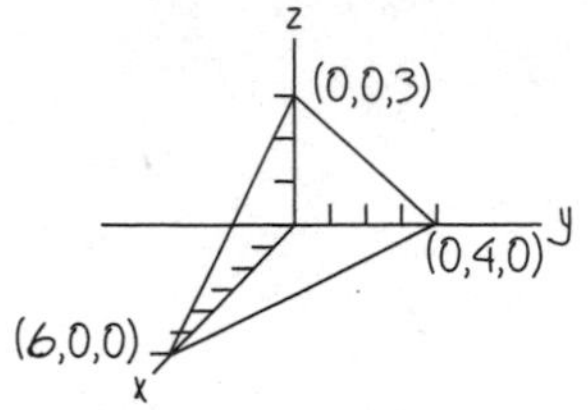

EXAMPLE 2—

$2x + 3y + 4z = 12$

We will find the three intercepts.

$x = 0$, $y = 0$. So $z = 3$. (0,0,3). Similarly, (0,4,0). Also, (6,0,0).

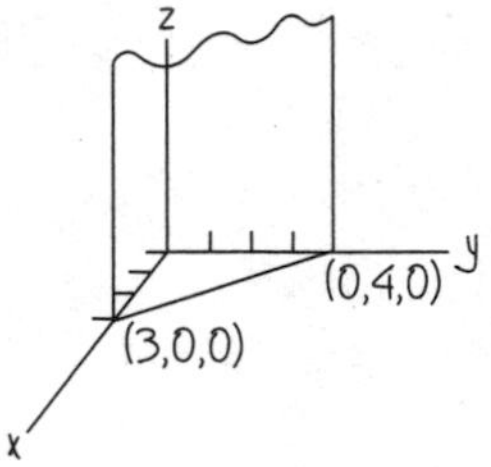

EXAMPLE 3—

$4x + 3y = 12$

Draw the line $4x + 3y = 12$.

z can be anything. See the figure.

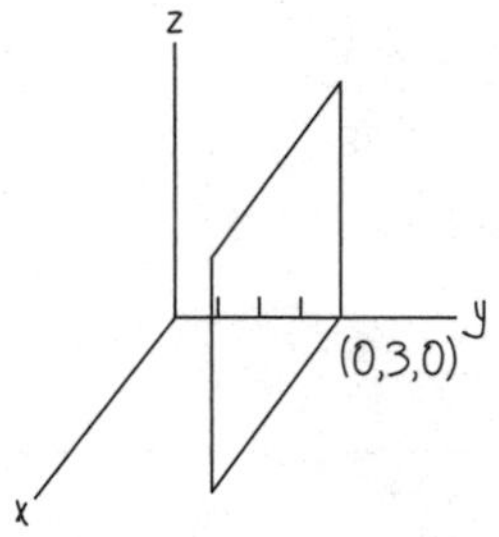

EXAMPLE 4—

$y = 3$

x and z can be anything (parallel to the x-z plane).

NOTE 1

$y = 3$ in 1D is a point; 2D is a line; 3D is a plane.

NOTE 2 (ACTUALLY 4,5,6).

$x = 0$ is y-z plane; $y = 0$ is x-z plane; $z = 0$ is x-y plane.

Introduction 2: Field

We are going to use a term that could have been used in elementary algebra, though hopefully not in great detail.

DEFINITION

Field—a set S with elements a, b, and c; operations + and ×; and:

1, 2. *Closure laws.* a + b is in S. ab is in S. (If you add or multiply, the answer is always in the set.)

NOTE

Odd integers are not closed under addition. If you add two odd integers, the answer is not odd.

3, 4. *Commutative laws.* a + b = b + a and ab = ba.

5, 6. *Associative laws.* (a + b) + c = a + (b + c) and (ab)c = a(bc).

7. *Identity for addition.* There is a number, call it 0, such that a + 0 = 0 + a = a.

8. *Identity for multiplication.* There is a number, call it 1, 1 ≠ 0, such that a(1) = 1(a) = a.

9. *Inverse for addition.* For every a, there is a number, call it –a, such that a + (–a) = (–a) + a = 0.

10. *Inverse for multiplication.* For every a ≠ 0, there is a number, call it 1/a such that a(1/a) = (1/a)a = 1.

11. *Distributive law.* a(b + c) = ab + ac.

That is, any set satisfying these 11 properties is a field. The two fields we will be dealing with are real numbers and complex numbers.

Introduction 3: Vectors in Pictures

A *vector* is an object with two properties: magnitude and direction. A *scalar* is an object with magnitude only.

EXAMPLE—

A wind blowing at 30 mph is a scalar. A wind blowing at 30 mph from the northwest is a vector. We will use letters at the end of the alphabet (**u**, **v**, and **w**) to indicate vectors. In many books, they are denoted with bold (dark) print. Letters at the beginning of the alphabet (a, b, and c) will be scalars.

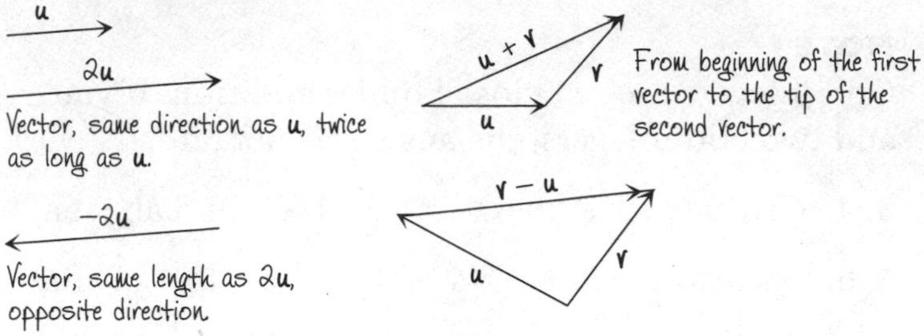
u
2u
Vector, same direction as u, twice as long as u.
−2u
Vector, same length as 2u, opposite direction
u + v
v
u
From beginning of the first vector to the tip of the second vector.
v − u
v
u

Enough introductions already. Let's get going. Let $\mathbf{u} = \langle u_1,u_2,u_3\rangle$ be a vector, indicated by angle brackets, from the origin (0,0,0) to the point (u_1,u_2,u_3).

EXAMPLE 1—

Suppose we have the vector $\mathbf{v} = \langle 3,5,7\rangle$. Find the vector AB from A(1,2,3) to B(4,7,10).

$\mathbf{AB} = \langle 4-1,\ 7-2,\ 10-3\rangle = \langle 3,5,7\rangle$.

NOTE

$\mathbf{v} = \mathbf{AB}$. Two vectors are *equal* (or *equivalent* in other books) if they have the same magnitude and direction. Pictorially, it means they have the same length and same direction.

Vector addition: $\mathbf{u} + \mathbf{v} = \langle u_1 + v_1,\ u_2 + v_2,\ u_3 + v_3\rangle$

Scalar multiplication: $c\mathbf{u} = \langle cu_1,cu_2,cu_3\rangle$

$\mathbf{0}$ vector $= \langle 0,0,0\rangle$: $-\mathbf{u} = \langle -u_1,-u_2,-u_3\rangle$

magnitude (length): $|\mathbf{u}| = (u_1^2 + u_2^2 + u_3^2)^{1/2}$ – 3D Pythag.

NOTE 1

In some books, magnitude is indicated by $\|\mathbf{u}\|$.

NOTE 2

Don't confuse magnitude with $|c|$, the absolute value of a scalar.

The set of vectors V with field F satisfies the following properties:

1. *Closure.* $\mathbf{u} + \mathbf{v}$ is a vector
2. *Associative.* $(\mathbf{u} + \mathbf{v}) + \mathbf{w} = \mathbf{u} + (\mathbf{v} + \mathbf{w})$
3. *Identity.* $\mathbf{0} + \mathbf{v} = \mathbf{v} + \mathbf{0} = \mathbf{v}$
4. *Inverse.* $\mathbf{v} + (-\mathbf{v}) = (-\mathbf{v}) + \mathbf{v} = \mathbf{0}$
5. *Commutative.* $\mathbf{u} + \mathbf{v} = \mathbf{v} + \mathbf{u}$
6. *"Weird" associative law.* $a(b\mathbf{v}) = (ab)\mathbf{v}$
7. *"Weird" distributive law.* $(a + b)\mathbf{v} = a\mathbf{v} + b\mathbf{v}$
8. *Second "not so weird" distributive law.* $a(\mathbf{v} + \mathbf{w}) = a\mathbf{v} + a\mathbf{w}$
9. *"Normal" identity.* $1\mathbf{v} = \mathbf{v}$

Let us prove one of these, let us say number 7. Most of you will do this stuff at some time in a course about linear algebra or perhaps modern algebra, the course most mathematicians love.

PROOF

$(a + b)\mathbf{v} = (a + b)\,\langle v_1,v_2,v_3\rangle = \langle (a + b)v_1,(a + b)v_2,(a + b)v_3\rangle$

$= \langle av_1 + bv_1,av_2 + bv_2,av_3 + bv_3\rangle$

$= \langle av_1,av_2,av_3\rangle + \langle bv_1,bv_2,bv_3\rangle = a\mathbf{v} + b\mathbf{v}$

plenty of notes.

NOTE 1
Those nine properties form a *vector space.*

NOTE 2
The "weirdness" involves associative and distributive laws that involve two different things: vectors and scalars.

NOTE 3
Note that in law 6, three of the multiplications involve scalar multiplication, while the fourth is scalar times a scalar. In 7, one plus sign adds vectors, and one plus sign adds scalars.

NOTE 4
Properties 1–4 form a group.

NOTE 5
Properties 1–5 form a commutative or Abelian group, named after the mathematician Abel.

NOTE 6
You will talk about all of these things in a linear algebra course.

Dot Product

$\mathbf{u} \cdot \mathbf{v} = u_1v_1 + u_2v_2 + u_3v_3$

NOTE
The dot product of two vectors is a scalar!!!!!!!!!

Properties of dot products

1. $\mathbf{u} \cdot \mathbf{u} = |\mathbf{u}|^2$
2. $\mathbf{u} \cdot \mathbf{v} = \mathbf{v} \cdot \mathbf{u}$
3. $\mathbf{u} \cdot (\mathbf{v} + \mathbf{w}) = \mathbf{u} \cdot \mathbf{v} + \mathbf{u} \cdot \mathbf{w}$
4. $(c\mathbf{u}) \cdot \mathbf{v} = c(\mathbf{u} \cdot \mathbf{v}) = \mathbf{u} \cdot (c\mathbf{v})$
5. $\mathbf{0} \cdot \mathbf{v} = 0$
6. *Important:* $\mathbf{u} \cdot \mathbf{v} = |\mathbf{u}|\ |\mathbf{v}| \cos \theta$, θ is angle between u and v
7. *Important:* **u**, **v** are perpendicular (ORTHOGONAL) if $\mathbf{u} \cdot \mathbf{v} = 0$
8. Scalar projection of u on v: $\text{proj}_{\mathbf{v}}\mathbf{u} = (\mathbf{u} \cdot \mathbf{v})/|\mathbf{v}|$

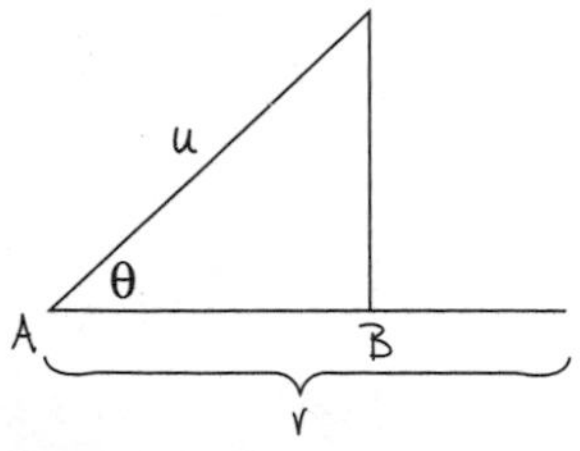

9. Vector projection of u on v in the direction of v: $\text{proj}_\mathbf{v}\mathbf{u} = (\mathbf{u} \cdot \mathbf{v}/|\mathbf{v}|) \div (\mathbf{v}/|\mathbf{v}|)$; $|AB|$ is $\text{proj}_\mathbf{v}\mathbf{u}$

NOTE

A *unit vector* $= \mathbf{v}/|\mathbf{v}|$.

Proofs of 1–9 are found in most calc books.

Let's finally do some examples. Even though the intro was long and tedious, the examples are mostly short and easy.

Given $\mathbf{u} = \langle 1,2,3\rangle \quad \mathbf{v} = \langle 4,-5,6\rangle \quad \mathbf{w} = \langle -2,-3,5\rangle$

EXAMPLE 1—

$3\mathbf{u} - 5\mathbf{v} \quad 3\langle 1,2,3\rangle - 5\langle 4,-5,6\rangle = \langle 3,6,9\rangle + \langle -20,25,-30\rangle$

$$= \langle -17,31,-21\rangle$$

EXAMPLE 2—

We can determine the angle between u and v. Since $\mathbf{u} \cdot \mathbf{v} = |\mathbf{u}|\ |\mathbf{v}| \cos\theta$, thennnnnnnn

$$\theta = \frac{\mathbf{u} \cdot \mathbf{v}}{|\mathbf{u}|\ |\mathbf{v}|} = \cos^{-1}\left(\frac{(1)(4) + 2(-5) + 3(6)}{\sqrt{14}\sqrt{77}}\right)$$

$$= \cos^{-1}\left(\frac{12}{\sqrt{14}\sqrt{77}}\right)$$

To find the angle, use a calculator.

EXAMPLE 3—

Find a vector of length 5 orthogonal to **w**.

A. *Orthogonal* means the dot product is 0.

$\mathbf{s} \cdot \mathbf{w} = 0$; $s_1w_1 + s_2w_2 + s_3w_3 = -2s_1 - 3s_2 + 5s_3 = 0$

The easiest way is to make one component 0, say s_3, and then reverse w_1 and w_2, changing one sign.

$\mathbf{s} = \langle 3,-2,0\rangle$; (See? $\mathbf{s} \cdot \mathbf{w} = 0$)

B. The length is not 5: first make it 1, a unit vector.

$$\mathbf{s}/|\mathbf{s}| = \frac{\langle 3,-2,0\rangle}{\sqrt{13}}$$

C. Lastly, multiply by 5 (or −5), since $1 \times 5 = 5$!

The answer is $\pm 5\langle 3,-2,0\rangle/\sqrt{13}$

EXAMPLE 4—

Find scalar and vector projection of v on u. (Be careful!)

A. $\text{proj}_{\mathbf{u}}\mathbf{v} = \mathbf{v} \cdot \mathbf{u}/|\mathbf{u}| = \dfrac{12}{\sqrt{14}}$

B. $\mathbf{proj_u v} = \left(\dfrac{\mathbf{v} \cdot \mathbf{u}}{|\mathbf{u}|}\right)\left(\dfrac{\mathbf{u}}{|\mathbf{u}|}\right) = 12/\sqrt{14}$

$(\langle 1,2,3\rangle/\sqrt{14}) = 6/7\langle 1,2,3\rangle$

Unfortunately, two more intros are needed, and then we get serious.

Alternate Vector Notation

Let $\mathbf{i}$ = vector of length 1 in positive x direction: $\langle 1,0,0\rangle$. Similarly $\mathbf{j} = \langle 0,1,0\rangle$ and $\mathbf{k} = \langle 0,0,1\rangle$. Thus, the vector $\langle 3,4,5\rangle$ can be written $3\mathbf{i} + 4\mathbf{j} + 5\mathbf{k}$.

Here, I evaluate a 3×3 determinant. This only works for 3×3. See a precalc or linear algebra book for more.

Recopy first two columns.
Down, add products.
Up, subtract products.

$$\begin{vmatrix} a & b & c \\ d & e & f \\ g & h & i \end{vmatrix}$$

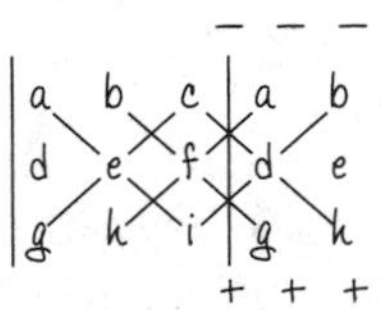

= aei + bfg + cdh − gec − hfa − idb

Cross Product

Finally, the last introduction!!! $\mathbf{u} \times \mathbf{v}$, read "u cross v," is the *cross product.* How do you get it?

Evaluate the following:

$$\begin{vmatrix} i & j & k \\ u_1 & u_2 & u_3 \\ v_1 & v_2 & v_3 \end{vmatrix}$$

What does it look like? Take your right hand; put it on the u in the figure; curl your fingers toward v; u × v is a *vector* in the direction of your thumb.

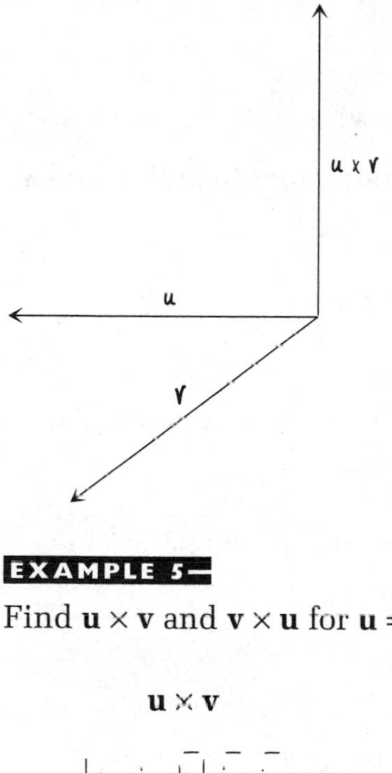

EXAMPLE 5—

Find **u** × **v** and **v** × **u** for **u** = ⟨1,2,3⟩, **v** = ⟨6,5,4⟩.

u × v

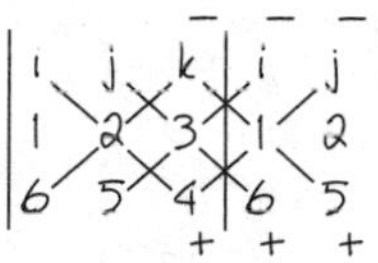

v × u

$$\left|\begin{array}{ccc|cc} i & j & k & i & j \\ 6 & 5 & 4 & 6 & 5 \\ 1 & 2 & 3 & 1 & 2 \end{array}\right.$$

i(2)(4) + j(3)(6) + k(1)(5)

−k(6)(2) − i(3)(5) − j(1)(4)

= −7i + 14j − 7k

i(5)(3) + j(4)(1) + k(6)(2)

−k(5)(1) − i(4)(2) − j(6)(3)

= 7i − 14j + 7k

Notice **u** × **v** = −**v** × **u**!!!!! This is the first of several rules.

Properties of Cross Products

1. $\mathbf{u} \times \mathbf{v} = -\mathbf{v} \times \mathbf{u}$. (The cross product is *anticommutative*).
2. $\mathbf{u} \times (\mathbf{v} + \mathbf{w}) = \mathbf{u} \times \mathbf{v} + \mathbf{u} \times \mathbf{w}$.
3. $\mathbf{u} \times \mathbf{0} = \mathbf{0} \times \mathbf{u} = \mathbf{0}$.
4. If $\mathbf{u} = c\mathbf{v}$, then $\mathbf{u} \times \mathbf{v} = \mathbf{0}$. (Cross product of parallel vectors is $\mathbf{0}$.)
5. $(\mathbf{u} \times \mathbf{v}) \cdot \mathbf{w} = \mathbf{u} \cdot (\mathbf{v} \times \mathbf{w})$.
6. $\mathbf{u} \times (c\mathbf{v}) = (c\mathbf{u}) \times \mathbf{v} = c(\mathbf{u} \times \mathbf{v})$.
7. *Important:* $\mathbf{u} \times \mathbf{v}$ is perpendicular to both $\mathbf{u}$ and $\mathbf{v}$.
8. $|\mathbf{u} \times \mathbf{v}| = |\mathbf{u}|\ |\mathbf{v}| \sin \theta$.

Again, the proofs are in most books.

NOTE 1
For emphasis, $\mathbf{u} \times \mathbf{v}$ is a vector; recall $\mathbf{u} \cdot \mathbf{v}$ is a scalar.

NOTE 2
By property 8, we can find the area of the parallelogram in the figure.

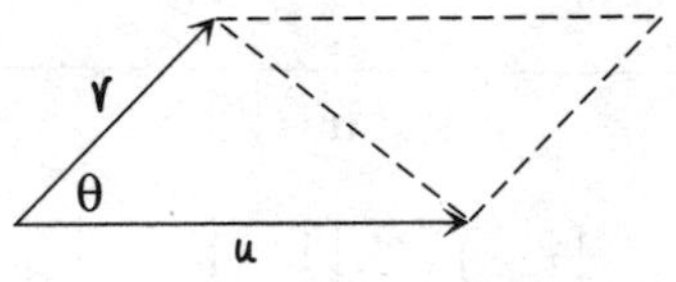

NOTE 3
By property 8, the area of the triangle would be ½ u × v!!!

NOTE 4
The cross product is only defined for 3D!!!!!!!!

Finally, we are ready to talk about planes and lines in 3D.

PLANES

Let **N** = Ai + Bj + Ck be a vector perpendicular to the plane. **N** is called the *normal.* Let $P_0(x_0,y_0,z_0)$ be the point where the normal hits the plane. If P(x,y,z) is any point in the plane, P_0P is perpendicular to **N**, since all vectors in the plane are.

$\mathbf{P_0P} = (x - x_0)i + (y - y_0)j + (z - z_0)k.$

Two vectors are perpendicular; therefore their dot product is 0.

$(Ai + Bj + Ck) \cdot ((x - x_0)i + (y - y_0)j + (z - z_0)k) = 0.$

We get.....

$A(x - x_0) + B(y - y_0) + C(z - z_0) = 0$

as the equation of the plane, or equivalently

$Ax + By + Cz = D$

LINES

We would like to find the equation of a line through the point $P_0(x_0,y_0,z_0)$ in the direction of $\mathbf{v} = \langle A,B,C\rangle$.

Let the position vector **r** = the vector from the origin to an arbitrary point on our line (x,y,z). The vector $\mathbf{r} = \langle x,y,x\rangle$. $\mathbf{r}_0$ = the vector from the origin to (x_0,y_0,z_0). $\mathbf{r}_0 = \langle x_0,y_0,z_0\rangle$.

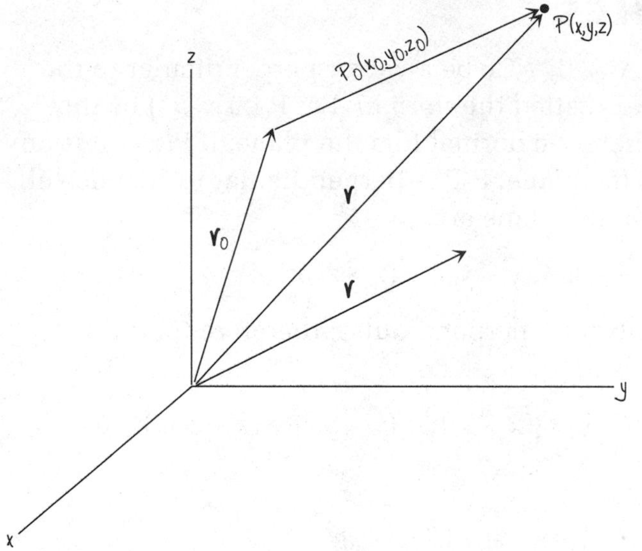

From the figure, $\mathbf{P_0P} = \mathbf{r} - \mathbf{r}$. Buuuuut, $\mathbf{P_0P}$ is parallel to $\mathbf{v}$. Soooo, $\mathbf{P_0P} = \mathbf{r} - \mathbf{r}_0 = t\mathbf{v}$, and t is a constant. The vector form is $\mathbf{r} = \mathbf{r}_0 + t\mathbf{v}$ or $\langle x,y,z\rangle = \langle x_0,y_0,z_0\rangle + t\langle A,B,C\rangle$.

Two vectors are equal if their components are equal, sooo we get the parametric form of the line.

$x = x_0 + At \qquad y = y_0 + Bt \qquad z = z_0 + Ct$

Solving for t, we get the symmetric form of the line . . .

$$\frac{x - x_0}{A} = \frac{y - y_0}{B} = \frac{z - z_0}{C}$$

NOTE

We will see more of $\mathbf{r}$, the position vector, later.

Let's finally do some real problems with lines and planes.

EXAMPLE 6—

Given points D(4,4,10), E(3,2,7), F(9,7,11)

A. Find the equation of the plane.

B. Find the area of triangle DEF.

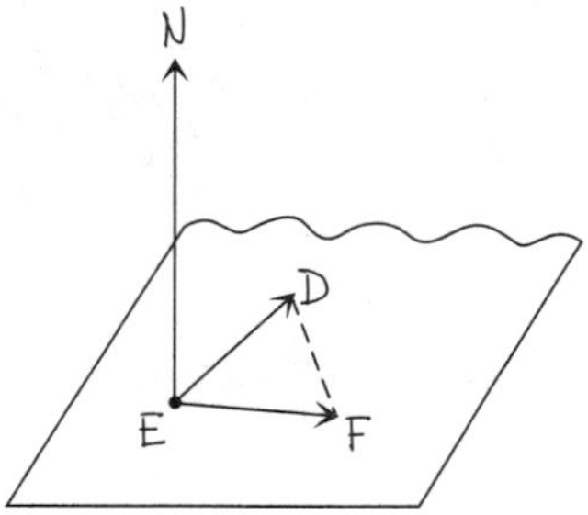

Let's do this first part the saane way. Three non-co-linear points determine a plane. We need to find the normal. We have the point (actually, three of them). Find the vectors **ED** and **EE**.

ED = ⟨1,2,3⟩; **EF** = ⟨6,5,4⟩; **N** = **ED** × **EF**

We did it before: **N** = ⟨–7,14,–7⟩.

Let's use $P_0 = (3,2,7)$. The plane is Ax + By + Cz = D = –7(3) + 14(2) + (–7)(7) = D = –42. The equation is –7x + 14y – 7z = –42 or x – 2y + z = 6.

Now for the second part. The area of a triangle is $\frac{1}{2}|\langle -7,14,-7\rangle| = 7\sqrt{6}/2$.

EXAMPLE 7A—

Find the line through points (4,5,6) and (7,9,14).
v = ⟨7 – 4,9 – 5,14 – 6⟩ = ⟨3,4,8⟩. Let $P_0 = (4,5,6)$.
Symmetrically, the line is

$$\frac{x - x_0}{A} = \frac{y - y_0}{B} = \frac{z - z_0}{C}$$

$$\frac{x - 4}{3} = \frac{y - 5}{4} = \frac{z - 6}{8}$$

If we set each equal to t, we get the parametric form

$x = 4 + 3t \qquad y = 5 + 4y \qquad z = 6 + 8t$

EXAMPLE 7B—

Find the line through (4,5,6) and (9,5,7). v = ⟨5,0,1⟩. Note the y component stays the same; y always equals 5. The symmetric form becomes

$$\frac{x - 4}{5} = \frac{z = 6}{1}, \; y = 5$$

EXAMPLE 8—

Find the equation of the plane perpendicular to the line (x – 4)/5 = (y + 3)/4 = (z – 2)/8 through the point (8,9,7).

This one's easy!!! If you were to draw the picture, you would see that the line is in the same direction as the normal. So v of the line is the N of the plane. $\mathbf{N} = \langle 5,4,8 \rangle$. Point (8,9,7).

$Ax + By + Cz = D = 5(8) + 4(9) + 8(7) = D = 132.$

The answer is $5x + 4y + 8z = 132$.

EXAMPLE 9—

Find the angle between $2x + 3y + 4z = 5$ and $6x - 2z = 11$.

In drawing the picture, the angle between the planes is the same as the angle between the normals.

$\mathbf{N}_1 = \langle 2,3,4 \rangle$ and $\mathbf{N}_2 = \langle 6,0,-2 \rangle$

$$\theta = \cos^{-1} \frac{\mathbf{N}_1 \cdot \mathbf{N}_2}{|\mathbf{N}_1|\ |\mathbf{N}_2|}$$

$$= \cos^{-1}\left(\frac{2(6) + 3(0) + 4(-2)}{\sqrt{(4+9+16)}\sqrt{(36+0+4)}}\right)$$

$$\theta = \cos^{-1} \frac{4}{\sqrt{29}\sqrt{40}}$$

EXAMPLE 10—

Find the equation of the line that is the intersection of the planes $2x + 3y + 4z = 14$ and $3x - 4z = 12$.

If you draw two planes that meet, you will find that the line of intersection is perpendicular to both normals. Therefore, the vector in the direction of the line v is in the same direction as $\mathbf{N}_1 \times \mathbf{N}_2$.

We need a point. We have two equations in three unknowns. We can arbitrarily assign one variable, say z = 0. Then x = 4 and y = 2. The point is (4,2,0).

$$\left|\begin{array}{ccc|cc} i & j & k & i & j \\ 2 & 3 & 4 & 2 & 3 \\ 3 & 0 & -4 & 3 & 0 \end{array}\right|$$

$\mathbf{v} = \langle -12,20,-9 \rangle$

The equation of the line is

$$\frac{x-4}{-12} = \frac{y-2}{20} = \frac{z-0}{-9}$$

EXAMPLE 11A—

Find the point where these two lines meet.

$$\frac{x-0}{1} = \frac{y-0}{-1} = \frac{z+6}{2}, \qquad \frac{x-1}{-1} = \frac{y-1}{3} = \frac{z-0}{2}$$

Put the lines in parametrics. Set the first equal to t and the second equal to s.

$x = t \quad y = -t \quad z = -6 + 2t, \qquad x = 1 - s \quad y = 1 + 3s \quad z = 0 + 2s.$

For the lines to meet, the x, y, and z numbers must match. Setting x's and y's equal, we get $t = 1 - s$ and $-t = 1 + 3s$. Solving, we get $t = 2$, $s = -1$. Substituting in both parametrics, we get $x = 2$, $y = -2$, and $z = -2$, which is the point where the lines meet.

EXAMPLE 11B—

Let's show what happens if they don't meet.

$$\frac{x-1}{2} = \frac{y-3}{2} = \frac{z-3}{6}, \qquad \frac{x}{1} = \frac{y}{2} = \frac{z-5}{3}$$

Parametrically, $x = 1 + 2t$, $y = 3 + 2t$, $z = 3 + 6t$; $x = s$, $y = 2s$, $z = 5 + 3s$. Setting x and y equal, $1 + 2t = s$, $3 + 2t = 2s$; $s = 2$, $t = \frac{1}{2}$.

$x = 1 + 2t = 1 + 2(\frac{1}{2}) = 2 \qquad x = s = 2$

$y = 3 + 2t = 3 + 2(\frac{1}{2}) = 4 \qquad y = 2s = 2(2) = 4$

$z = 3 + 6t = 3 + 6(\frac{1}{2}) = 6 \qquad z = 5 + 3s = 5 + 3(2) = 11$

The z's are not the same; therefore, the lines don't meet. Oh, well . . . let's go to the next problem.

EXAMPLE 12—

Find the distance from (1,4,7) to $2x + 3y + 5z = 100$.
The distance from point $P_0(x_0,y_0,z_0)$ to plane $Ax + By + Cz = D$ is given by

$$\frac{|Ax_0 + By_0 + Cz_0 - D|}{\sqrt{(A^2 + B^2 + C^2)}} = \frac{|2(1) + 3(4) + 5(7) - 100|}{\sqrt{(2^2 + 3^2 + 5^2)}}$$

$$= \frac{51}{\sqrt{38}}$$

NOTE 1

To find this formula, find a point on the plane. $Q = (0,0,D/C)$ will work in most cases. Find $\mathbf{P_0Q}$.

The distance formula is the absolute value of the scalar projection of $\mathbf{P_0Q}$ on the normal $\mathbf{N}$.

NOTE 2

This is very similar to the distance from a point (x_0,y_0) to a line $Ax + By = C$ in the plane

$$d = \frac{|Ax_0 + By_0 - C|}{\sqrt{A^2 + B^2}}$$

EXAMPLE 13—

Find the distance between lines.

"Distance" means perpendicular distance. Perpendicular to both lines is the cross product $\mathbf{v}_1 \times \mathbf{v}_2 = \mathbf{N}$, where v_1 and v_2 are the directions of each line. Find a point P on line 1 and point Q on line 2. Form **PQ**. Find the absolute value of the scalar projection of **PQ** on **N**.

PARAMETERS A LA 3D (OR MORE)

In *Calc II,* we talked about parameters. We would like to talk about parameters again in three (or more) dimensions, but we will do it in a more elegant form. Consider the points (x,y,z), which are all functions of t. The x(t)i + y(t)j + z(t)k is the vector from the origin to (x,y,z). We get the position vector **r** = ⟨x(t),y(t),z(t)⟩.

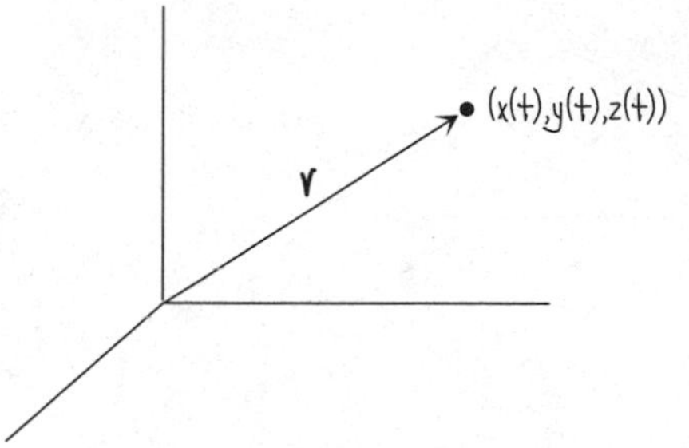

If we follow the tip of **r**, we will get the 3D curve it generates.

As in *Calc II,* the velocity vector **v** = ⟨x′(t),y′(t),z′(t)⟩. The acceleration **a** = ⟨x″(t),y″(t),z″(t)⟩. The speed $|\mathbf{v(t)}| = \sqrt{[x'(t)^2 + y'(t)^2 + z'(t)^2]}$.

Aaaaaand the arc length

$$s = \int_{t_1}^{t_2} \sqrt{[x'(t)^2 + y'(t)^2 + z'(t)^2]^{1/2}}\, dt.$$

Let us give an example.

x = 2 cos t,   y = 2 sin t,   z = t

If you look at x and y only, we get the circle $x^2 + y^2 = 4$. But z = t means the point moves up also. We get a circular spiral or *helix.*

NOTE

The path of the earth around the sun is an elliptical helix.

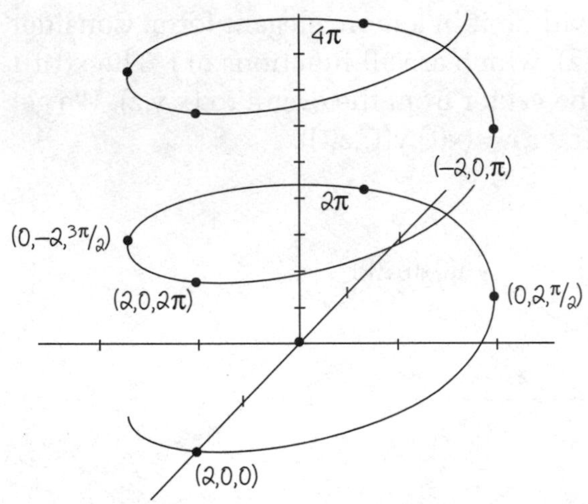

Let us do the problem. Find **r**, **v**, **a**, speed, and arc length from 0 to 2π.

$\mathbf{r} = \langle 2 \cos t, 2 \sin t, t\rangle$; $\mathbf{v} = \langle -2 \sin t, 2 \cos t, 1\rangle$;

$\mathbf{a} = \langle -2 \cos t, -2 \sin t, 0\rangle$

$|\mathbf{v}| = \sqrt{[(-2 \sin t)^2 + (2 \cos t)^2 + 1^2]}$;

$s = \int_0^{2\pi} \sqrt{3}\, dt = 2\pi\sqrt{3}$

CHAPTER 2

CURVES IN 3D: CONES, CYLINDERS, QUADRICS

SKETCHING IN 3D

We will do a little drawing in three dimensions. At least for me, this is very difficult. Before I teach this section in class, I practice for hours and hours. Fortunately, much of this can now be done on computers.

Cones

I'll bet you think you know what a cone is. Let us formally define a cone.

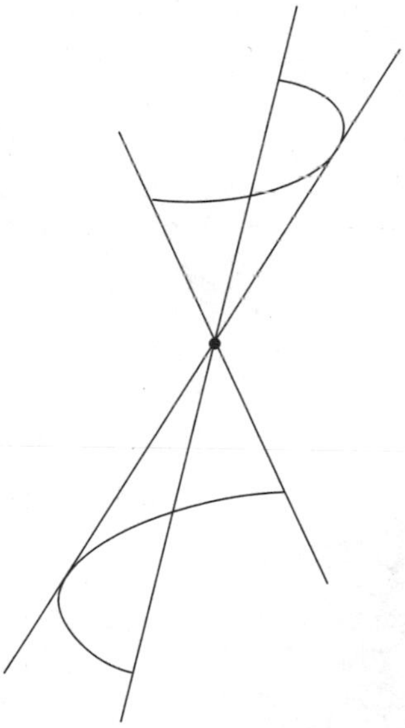

DEFINITION

Cone—Given a curve in a plane, take a point not in the plane of the curve. All lines through that curve and the point form a cone.

You might say that this is not the cone you know. The cone you know is not infinite; it is cut at the top and at the point (truncated). If you pour water in it, the water wouldn't fall out (it's closed). The base is circular. The height, if you could draw it, is not to the middle of the base (right). So the cone we are used to, the one we eat with ice cream, is a truncated, closed, circular, right cone. But don't ask for one at the soda fountain.

Cylinders

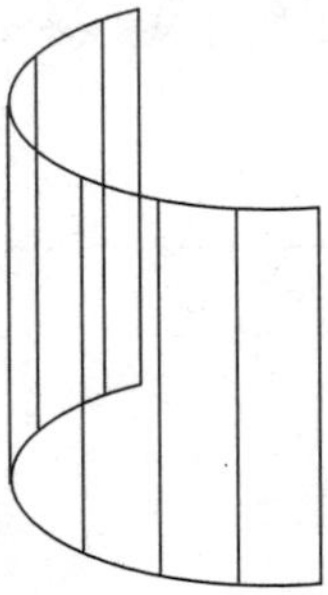

Again, the general definition of a cylinder is different.

DEFINITION

Cylinder—Given a curve in a plane, take a line L through the curve. Draw all possible lines through the curve parallel to L; you have a cylinder.

NOTE

A plane is a cylinder. It is the set of all lines through a given line.

Quadric Surfaces

Let us sketch a couple.

ellipsoid

$$\frac{x^2}{9} + \frac{y^2}{16} + \frac{z^2}{9} = 1$$

We will do this by sketching the "main plane" traces.

y-z plane, x = 0:

$$\frac{y^2}{16} + \frac{z^2}{9} = 1, \text{ ellipse}$$

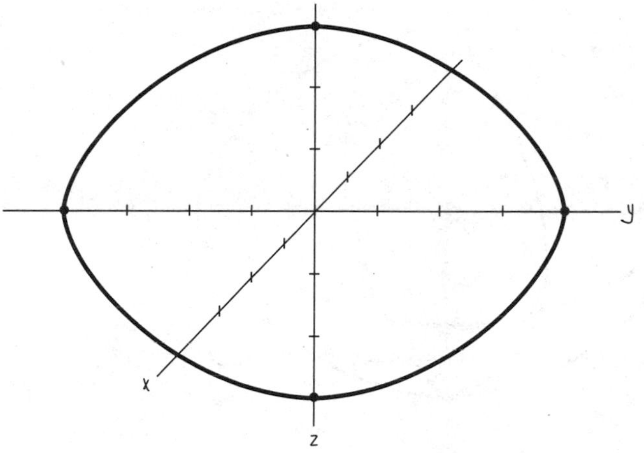

x-z plane, y = 0:

$$\frac{x^2}{9} + \frac{z^2}{9} = 1, \text{ circle}$$

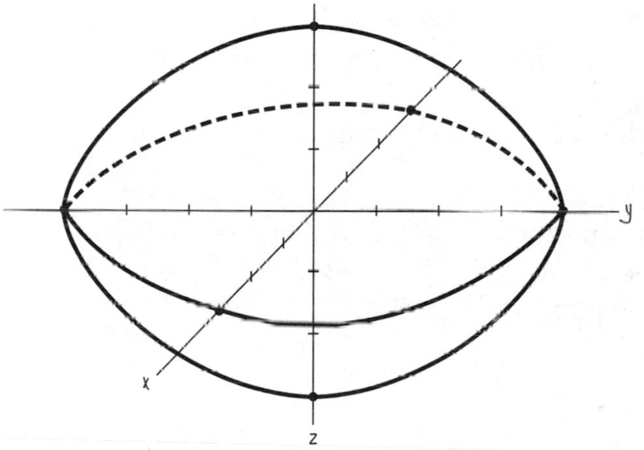

aaaaand x-y plane, z = 0:

$$\frac{x^2}{9} + \frac{y^2}{16} = 1, \text{ ellipse}$$

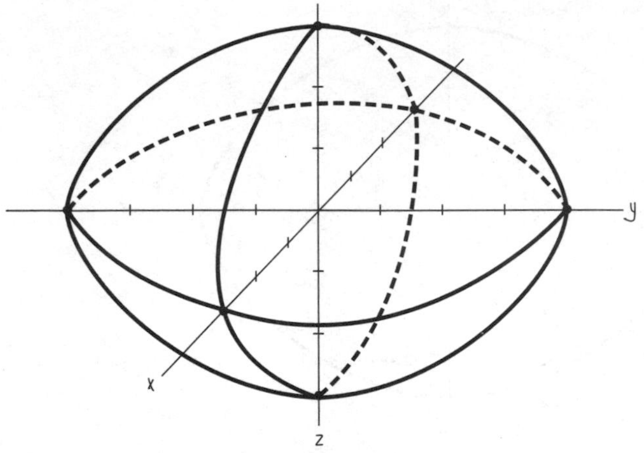

We get a pretty good idea of the ellipsoid, but it does take practice, especially if you see these pictures.

Let's do a circular paraboloid.

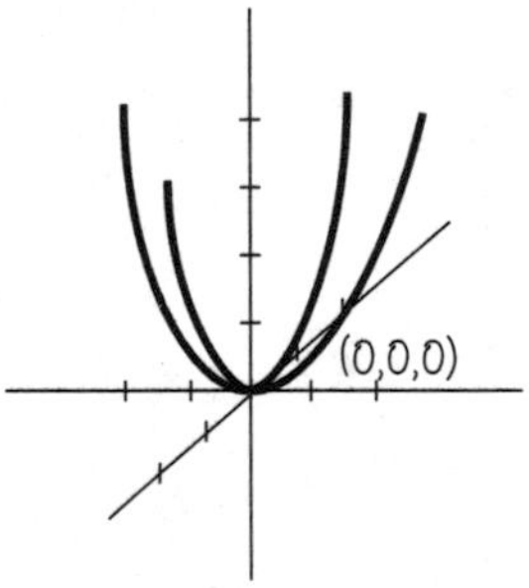

$z = x^2 + y^2$

If we let $z = 0$, we get the point (0,0,0). If we let $x = 0$, we get the parabola $z = y^2$. If we let $y = 0$, we get the parabola $z = x^2$.

Sketching this, we do not quite get a good idea.

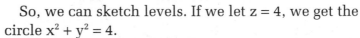

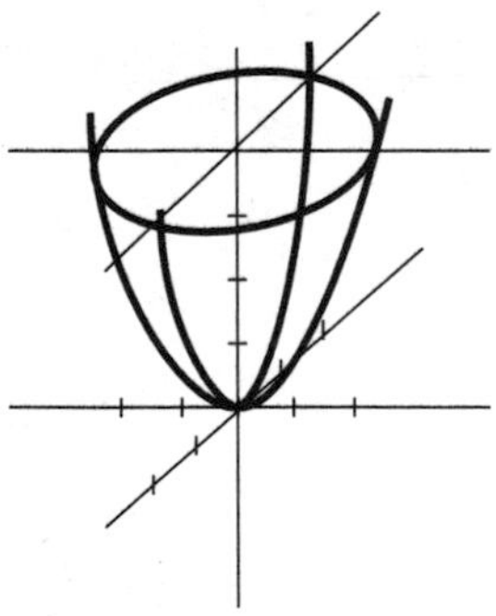

So, we can sketch levels. If we let $z = 4$, we get the circle $x^2 + y^2 = 4$.

Now the picture is clearer!

If we want a purely two-dimensional picture, we can graph levels only. This is much the same as a topographical map in geography, one that shows the heights of mountains, valleys, and so on.

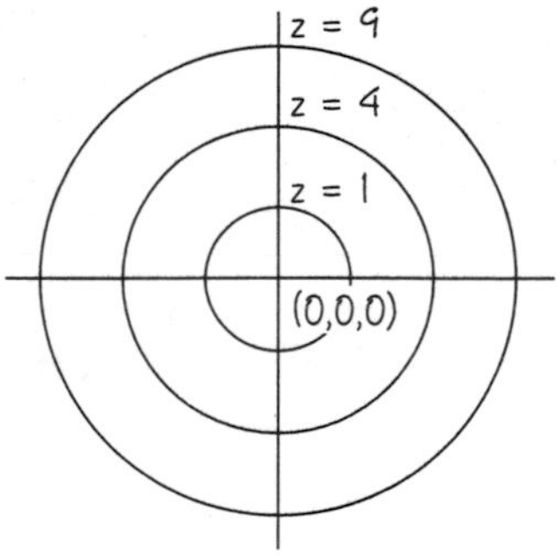

$z = 0$ We get the point (0,0,0)

$z = 1$ $x^2 + y^2 = 1$

$z = 4$ $x^2 + y^2 = 4$

$z = 9$ $x^2 + y^2 = 9$

We get the figure shown. You must imagine the height. For more of these pictures, see your local calculus books.

This naturally brings us to the next topic, a function of two variables. Just like in elementary algebra, we can define $f(x,y) = z$. Given a set D (domain) of ordered pairs (x,y), we assign to each ordered pair exactly one real number.

NOTE 1

The range is a real number.

NOTE 2

The domain could be, for example, an ordered triplet.

EXAMPLE

$f(x,y) = x^2 + y^2$; $f(3,5) = 3^2 + 5^2 = 34$

The ordered triple would be (3,5,34).

NOTE

This is another way of putting the circular paraboloid.

CHAPTER 3

LIMITS IN 1D AND 2D

LIMITS AND CONTINUITY

At this point, before I do anything else, I go back to my *Calc I* and explain the technical definition of a limit in detail. It usually occurs early in Calc I, before anyone is ready for it. Now you are ready. (Most people understand the nontechnical idea of a limit.) Then we will talk about nontechnical and technical definitions of limits in two variables.

FORMAL DEFINITION

We will now tackle the most difficult part of basic calculus, the theoretical definition of *limit.* It took two of the finest mathematicians of all time, Newton and Leibniz, to first formalize this topic. It is *not* essential to the rest of basic calculus to understand this definition. It is hoped that this explanation will give you some understanding of how really amazing calculus is and how brilliant Newton and Leibniz must have been. Remember, this is an approximating process that many times gives exact or very, very close answers. To me, this is mind-boggling, terrific, stupendous, unbelievable, awesome, cool, and every other great word you can think of.

DEFINITION

$\lim_{x \to a} f(x) = L$ if and only if, given an $\varepsilon > 0$, there exists a $\delta > 0$ such that if $0 < |x - a| < \delta$, then $|f(x) - L| < \varepsilon$.

NOTE

ε = epsilon and δ = delta, two letters of the Greek alphabet.

TRANSLATION 1

Given ε, a small positive number, we can always find a δ, another small positive number, such that if x is within a distance δ from a but not exactly at a, then f(x) is within a distance ε from L.

TRANSLATION 2

We will explain this definition using an incorrect picture. I feel this gives you a much better idea than the correct picture, which we will do next.

Interpret $|x - a|$ as the distance between x and a, but instead of the one-dimensional picture it really is, imagine that there is a circle around the point a of radius δ. $|x - a| < \delta$ stands for all x values that are inside this circle. Similarly, imagine a circle of radius ε around L, with $|f(x) - L| < \varepsilon$ being the set of all points f(x) that are inside this circle.

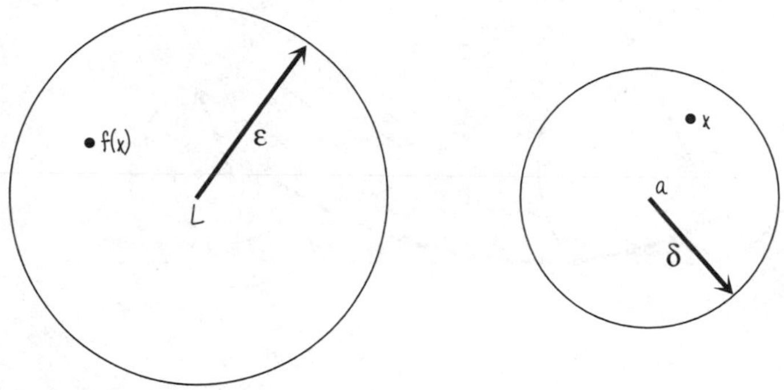

The definition says that, given an $\varepsilon > 0$ (given a circle of radius ε around L), we can find a $\delta > 0$ (circle of radius δ around a) such that, if $0 < |x - a| < \delta$ (if we take any x inside this circle), then $|f(x) - L| < \varepsilon$, [f(x)] will be inside the circle of radius ε but not exactly at L.

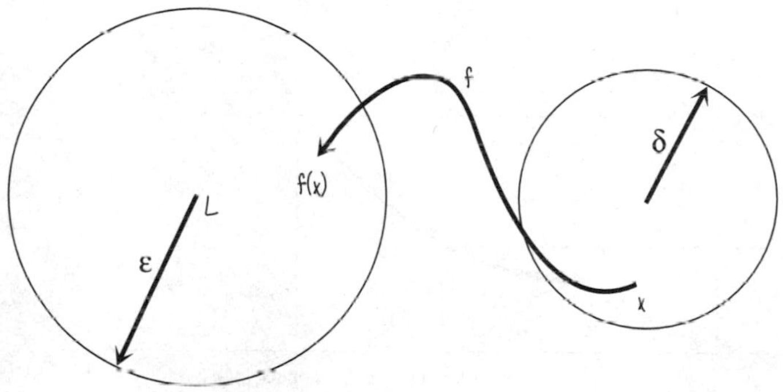

Now take another epsilon ε_2, positive but smaller than ε (a smaller circle around L); there exists another delta δ_2, usually a smaller circle around a, such that if $0 < |x - a| < \delta_2$, then $|f(x) - L| < \varepsilon_2$.

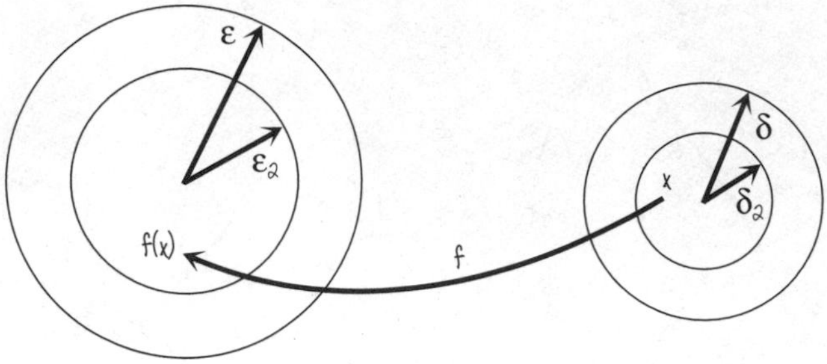

Now take smaller and smaller positive epsilons; we can find smaller and smaller deltas. In the limit, as the x circle shrinks to a, the f(x) circle shrinks to L. Read this a number of times!!!

TRANSLATION 3

Let us see the real picture. $y = f(x)$. $|x - a| < \delta$ means $a - \delta < x < a + \delta$. $|y - L| < \varepsilon$ means $L - \varepsilon < y < L + \varepsilon$.

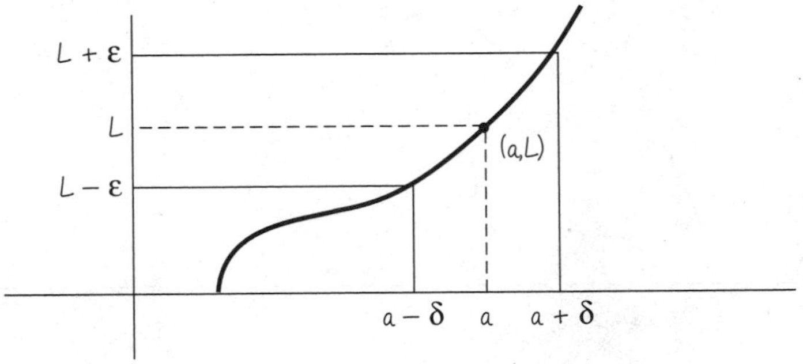

Given an $\varepsilon > 0$, if we take any x value such that $0 < |x - a| < \delta$, the interval on the x axis, and find the corresponding $y = f(x)$ value, this y value must be within ε of L; that is, $|f(x) - L| < \varepsilon$.

Take a smaller ε_2. We can find a δ_2 such that $0 < |x - a| < \delta_2$, $|f(x) - L| < \varepsilon_2$. The smaller the epsilon, the smaller the delta. $y = f(x)$ goes to L as x goes to a.

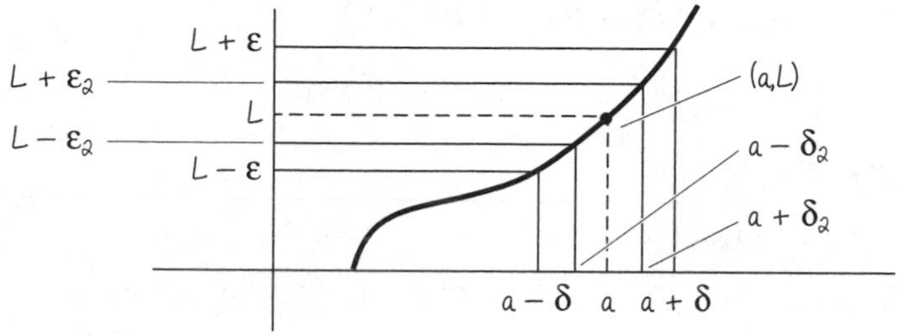

NOTE
The point (a,L) may or may not actually be there, depending on the function. But remember, we are only interested in values of x very, very close to a but not exactly at a.

Although this definition is extremely difficult, its application is pretty easy. We need to review six facts: four about absolute value and two about fractions.

1. $|ab| = |a|\,|b|$.
2. $\left|\dfrac{a}{b}\right| = \dfrac{|a|}{|b|}$.
3. $|a - b| = |b - a|$.
4. $|a + b| \le |a| + |b|$.
5. In comparing two positive fractions, if the bottoms are the same, and both numerators and denominators are positive, the larger the top and the larger the fraction. $2/7 < 3/7$.
6. If the tops are the same, the larger the bottom and the smaller the fraction. $3/10 > 3/11$.

Now let us do some problems.

EXAMPLE 1—

Using ε, δ, prove $\lim_{x\to 2} (4x - 3) = 5$.

In the definition $\lim_{x\to a} f(x) = L$, $f(x) = 4x - 3$, $a = 2$, $L = 5$. Given an $\varepsilon > 0$, we must find a $\delta > 0$, such that if $0 < |x - 2| < \delta$, $|(4x - 3) - 5| < \varepsilon$.

$$|(4x - 3) - 5| = |4x - 8| = |4(x - 2)|$$
$$= |4|\ |x - 2| < 4 \cdot \delta = \varepsilon \qquad \delta = \varepsilon/4$$

EXAMPLE 2—

Prove $\lim_{x\to 4} (x^2 + 2x) = 24$.

Given an $\varepsilon > 0$, we must find a $\delta > 0$ such that, if $0 < |x - 4| < \delta$, $|x^2 + 2x - 24| < \varepsilon$.

$$|x^2 + 2x - 24| = |(x + 6)(x - 4)| = |x + 6|\ |x - 4|$$

We must make sure that $|x + 6|$ does not get too big. We must always find a delta, no matter how small. We must take a preliminary $\delta = 1$. $|x - 4| < 1$, which means $-1 < x - 4 < 1$ or $3 < x < 5$. In any case, $x < 5$. Sooooo . . .

$$|x + 6| \le |x| + |6| < 5 + 6 = 11$$

Finishing our problem, $|x + 6|\ |x - 4| < 11 \cdot \delta = \varepsilon$. $\delta =$ minimum $(1, \varepsilon/11)$.

EXAMPLE 3—

Prove $\lim\limits_{x\to 5} \dfrac{2}{x} = \dfrac{2}{5}$.

$$\left|\frac{2}{x} - \frac{2}{5}\right| = \left|\frac{10 - 2x}{5x}\right| = \frac{|2(5 - x)|}{|5x|} = \frac{|2|\ |5 - x|}{|5|\ |x|}$$
$$= \frac{2|x - 5|}{5|x|}$$

Again take a preliminary delta $= 1$. $|x - 5| < 1$. So $4 < x < 6$. To make a fraction larger, make the top larger and the bottom smaller. $0 < |x - 5| < \delta$. We substitute δ on the top. Since $x > 4$, we substitute 4 on the bottom.

$$\frac{2|x - 5|}{5|x|} < \frac{2 \cdot \delta}{5 \cdot 4} = \frac{\delta}{10} = \varepsilon \qquad \delta = 10\varepsilon$$

If $\delta =$ minimum $(1, 10\varepsilon)$, $|2/x - 2/5| < \varepsilon$.

We are now ready for limits in two variables.

DEFINITION

$\lim_{(x,y)\to(a,b)} f(x,y) = L.$

Given an $\varepsilon > 0$, there exists a $\delta > 0$, such that if

$0 < |x - a| < \delta$ and $0 < |y - b| < \delta$, then $|f(x,y) - L| < \varepsilon$.

Nontechnically, the closer that (x,y) gets to the point (a,b), the closer that f(x,y) gets to L. In one variable, this procedure is usually quite easy. You take limits from the left and limits from the right. If both are the same, then the common number is the limit. (x,y) is in two dimensions. Therefore, there are an infinite number of paths into the point (a,b). Let us give two examples to show the complexity.

EXAMPLE 4—

Show that $\lim_{(x,y)\to(0,0)} \dfrac{xy}{x^2 + y^2}$ does not exist.

We will show that different paths to the origin give different limits. The limit along the y axis

$$\lim_{(0,y)\to(0,0)} \frac{xy}{x^2 + y^2} = 0/y^2 = 0.$$

The same is true on the x axis, the points (x,0) approaching (0,0). However, if we take the path y = x, (x,x), into the origin, we get

$$\lim_{(x,y)\to(0,0)} \frac{xy}{x^2 + y^2} = \frac{x^2}{x^2 + x^2} = 1/2.$$

Since the limits on two different paths are not the same, the limit does not exist!!!!!!!

It can get even trickier.

EXAMPLE 5—

$$\lim_{(x,y)\to(0,0)} \frac{x^2y}{x^4 + y^2} \text{ doesn't exist.}$$

On the x axis,

$$\lim_{(x,0)\to(0,0)} \frac{x^2y}{x^4+y^2} = 0/y^2 = 0$$

Similarly, on the y axis—and, as a matter of fact, on every line into the origin.....

$$y = kx, \qquad \lim_{(x,kx)\to(0,0)} \frac{x^2y}{x^4+y^2} = \frac{x^2(kx)}{x^4+(kx)^2} = \frac{kx}{x^2+k^2} = 0$$

However, if we take a parabolic path $y = x^2$ into the origin, we getttt

$$\lim_{(x,x^2)\to(0,0)} \frac{x^2y}{x^4+y^2} = \frac{x^2(x^2)}{x^4+(x^2)^2} = \frac{x^4}{x^4+x^4} = 1/2.$$

Since the limits on different paths vary, the limit at (0,0) does not exist.

NOTE

If you draw $0 < |x - a| < \delta$ and $0 < |y - b| < \delta$, it is a square around the point (a,b). In some books, the definition involves a circle, $\sqrt{(x-a)^2 + (y-b)^2} < \delta$. In doing problems with δ, ε, our definition is easier to use, much easier.

EXAMPLE 6—

Using δ, ε, prove $\lim_{(x,y)\to(2,3)} 4x + 5y = 23$.

Since as $x\to2$, $4x\to8$; and as $y\to3$, $5y\to15$

$$|4x + 5y - 23| = |4x - 8 + 5y - 15|$$

$$\le |4(x-2)| + |5(y-3)|$$

$$= 4|(x-2)| + 5|(y-3)|$$

$$< 4\delta + 5\delta = 9\delta = \varepsilon.$$

If $\delta = \varepsilon/9$, then $|4x + 5y - 23| < \varepsilon$.

If you note, the algebra, as compared to that from Calc I, is trickier. Here's another.

EXAMPLE 7—

Prove $\lim_{(x,y)\to(3,2)} x^2y = 18$.

$|x^2y - 18| = |x^2y - 2x^2 + 2x^2 - 18|$

We added and subtracted 0 in the form of $-2x^2 + 2x^2$ so that if you factored out the x^2, we would get $y - 2$, the delta. We could have subtracted and added 9y. It also works. Try it!!!!

$$= |x^2(y - 2) + 2(x + 3)(x - 3)|$$

$$< x^2\delta + 2|x + 3|\delta$$

Just like in Calc I, we take a preliminary $\delta = 1$, so the expression $|x + 3|$ doesn't get too large (and x^2 doesn't become large, either). $|x - 3| < 1$, then $-1 < x - 3 < 1$ or $2 < x < 4$, so $|x + 3| < 7$ (and $x^2 < 16$). Therefore,

$x^2\delta + 2|x + 3|\delta < 16\delta + (2)(7)\delta = 30\delta = \varepsilon$

If δ = minimum of $(1, \varepsilon/30)$, $|x^2y - 18| < \varepsilon$.

Mathematics is done in levels. You have gone through one level, which is covered in the Calculus I and II courses. Now you are beginning the second level. (There are many more.) If you recall Calc I, you remember that the next topic is derivatives. In 3D, this becomes a little more complicated.

The first thing we will talk about is partial derivatives, when function $z = f(x,y)$. We would like to take the derivative with respect to x and y. Let us see what that means.

PARTIAL DERIVATIVE

Take a curve in space, $z = f(x,y)$.

Make a slice with a knife, parallel to the y-z plane. In other words, x is a constant. There is a two-dimensional curve that forms with x constant. So we would like to know how z changes with respect to y, holding x a constant.

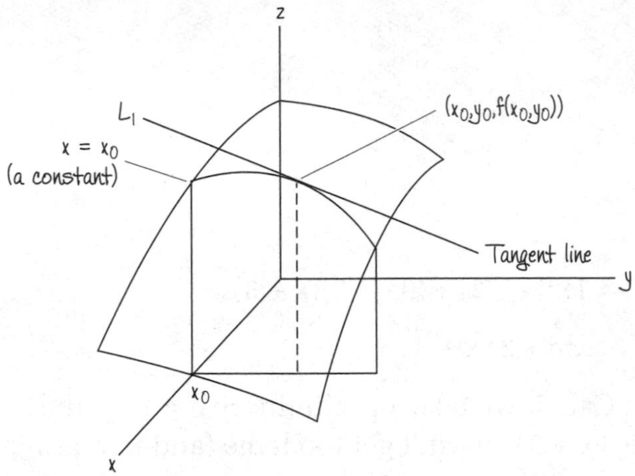

DEFINITION

The first partial derivative of f, with respect to y:

$$f_y = \lim_{\Delta y \to 0} \frac{f(x,y + \Delta y) - f(x,y)}{\Delta y}$$

Similarly, the first partial derivative of f, with respect to x:

$$f_x = \lim_{\Delta x \to 0} \frac{f(x + \Delta x,y) - f(x,y)}{\Delta x}$$

Alternate notations:

$$f_x = \frac{\partial f}{\partial x} = f_1$$

$$f_y = \frac{\partial f}{\partial y} = f_2$$

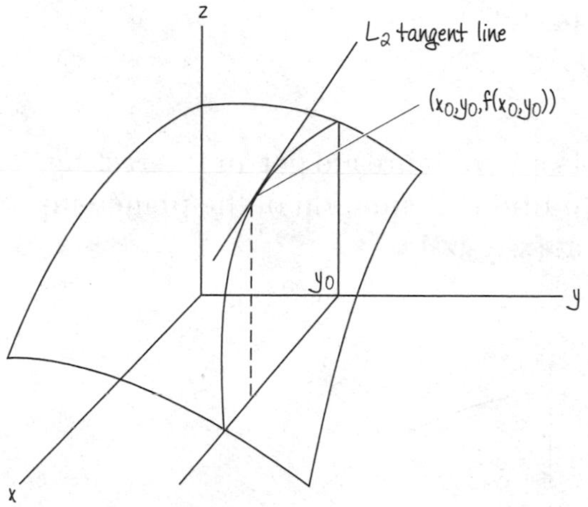

The doing of this is rather easy.

EXAMPLE 8—

Find f_{xxy} if $f(x,y) = x^5y^4 + x^6 + y^{10}$

$\left(\text{alternate notation: } \frac{\partial^3 f}{\partial y \partial^2 x}\right)$.

We read this left to right: begin with the first partial with respect to x and read right to left, then the partial with respect to x again, then the partial with respect to y.

$$f_x = \frac{\partial f}{\partial x} = 5x^4y^4 + 6x^5 + 0$$

Remember y is a constant.

$$f_{xx} = \frac{\partial^2 f}{\partial x^2} = 20x^3y^4 + 30x^4$$

y a constant again.

$$f_{xxy} = \frac{\partial^3 f}{\partial y \partial^2 x} = 80x^3y^3 + 0 = 80x^3y^3$$

x is the constant now.

We read this notation right to left!

EXAMPLE 9—

Find f_x and f_y if $f(x,y) = x^5 \tan(x^3y^4)$.

$$f_x = x^5 \sec^2(x^3y^4)\,[3x^2y^4] + \tan(x^3y^4)\,[5x^4]$$

The product rule—hold y a constant.

No product rule; remember, y is the variable, and x is the constant.

$f_y = x^5 \sec^2 (x^3y^4) [4x^3y^3]$

We combine the last two figures to get an interesting result. We wish to find the equation of the plane tangent to $z = f(x,y)$ at (x_0,y_0,z_0).

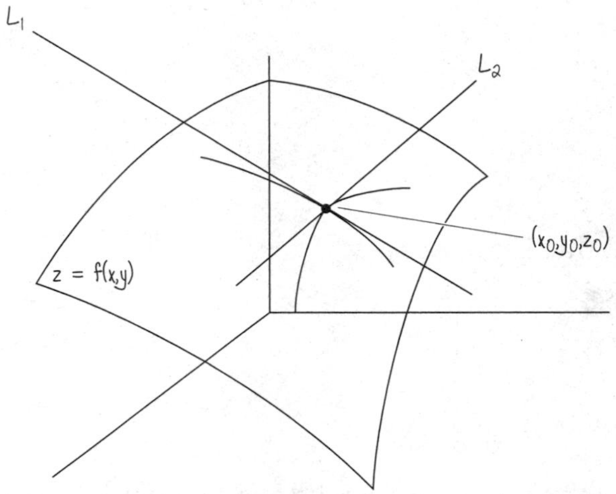

The tangent plane contains the lines L_1 and L_2. We have the point, namely (x_0,y_0,z_0) where $z_0 = f(x_0,y_0)$. We need the normal.

The direction of L_1, $\mathbf{v}_1 = 0i + 1j + f_yk$. Remember, there is no change in x; the slope is f_y $(z_y = f_y)$.

The direction of L_2, $\mathbf{v}_2 = 1i + 0j + f_xk$.

The normal:

$$\mathbf{N} = \mathbf{v}_1 \times \mathbf{v}_2 = \begin{vmatrix} i & j & j \\ 0 & 1 & f_y \\ 1 & 0 & f_x \end{vmatrix} = f_xi + f_yj - k = \mathbf{N}.$$

The equation of the tangent plane would be, in general,

$$A(x - x_0) + B(y - y_0) + C(z - z_0) = 0$$

In this case, $A = f_x$, $B = f_y$, $C = -1$.

$f_x(x - x_0) + f_y(y - y_0) - 1(z - z_0) = 0$

ooorrrrrrr

$z - z_0 = f_x(x_0,y_0)(x - x_0) + f_y(x_0,y_0)(y - y_0)$

We will get back to this a little later.

If you remember back to *Calc I,* in the first real chapter of calculus, we talked about the tangent line, the differential, and the chain rule.

Now that we have the partial derivatives (the derivative in the x and y direction), we would like to have the derivative in all directions and extend the differential, the chain rule, and the tangent plane. But first, we have to talk about the total change in $u = f(x,y)$, Δu. We use the letter u because we will extend to $u = f(x,y,z)$.

TOTAL INCREMENT

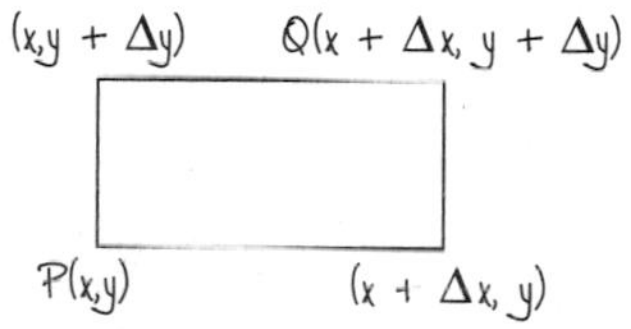

We are given the following:

1. $u = f(x,y)$ is defined in D.
2. f_x, f_y continuous in D.
3. The whole rectangle is in D.

We conclude that there exists numbers $\varepsilon_1,\varepsilon_2$ such that

$$\Delta u = f(x + \Delta x, y + \Delta y) - f(x,y)$$

$$= f_x\Delta x + f_y\Delta y + \varepsilon_1\Delta x + \varepsilon_2\Delta y$$

where both epsilons go to zero as the deltas go to zero.

PROOF

$$\Delta u = f(x + \Delta x, y + \Delta y) - f(x,y)$$

$$\Delta u = [f(x + \Delta x, y + \Delta y) - f(x, y + \Delta y)] + [f(x, y + \Delta y) - f(x,y)]$$

We add and subtract the term and get an interesting result. The first bracket has constant y. We can apply the mean value theorem. There exists a point w_1, where w_1 is between x and $x + \Delta x$, such that $f(x + \Delta x, y + \Delta y) - f(x, y + \Delta y) = f_x(w_1, y + \Delta y)\Delta x$.

NOTE

Recall the mean value theorem: f(x) is continuous on [a,b]; f′(x) exists on (a,b). There exists a point w between a and b, such that $f'(w) = f(b) - f(a)/(b - a)$. Multiplying through, we get $f(b) - f(a) = f'(w)(b - a)$. Applying here, we see that $f'(w) = f_x(w_1, y + \Delta y) \cdot \Delta x$ and $a = x$, $b = x + \Delta x$.

Similarly, the second bracket becomes $f(x, y + \Delta y) - f(x,y) = f_y(x, w_2)\Delta y$, where w_2 is between y and $y + \Delta y$.

Therefore, $\Delta u = f_x(w_1, y + \Delta y)\Delta x + f_y(x, w_2)\Delta y$.

If P and Q are very, very close, $f_x(w_1, y + \Delta y) = f_x(x,y) + \varepsilon_1$ and $f_y(x, w_2) = f_y(x,y) + \varepsilon_2$, where the epsilons go to 0 as the deltas go to 0. (English translation: if P and Q are close, $f_x(w_1, y + \Delta y) = f_x(x,y)$ plus a little, since f_x is continuous.)

Substituting into the previous equation, we get what we want, namely

$$\Delta u = f_x(x,y)\Delta x + f_y(x,y) + \varepsilon_1 \Delta x + \varepsilon_2 \Delta y$$

NOTE 1

We can define the *differential:* $du = f_x\, dx + f_y\, dy$.

NOTE 2

We can extend the last two items to three (or more) variables.

$$\Delta u = f_x \Delta x + f_y \Delta y + f_z \Delta z + \varepsilon_1 \Delta x + \varepsilon_2 \Delta y + \varepsilon_3 \Delta z$$

$$du = f_x\, dx + f_y\, dy + f_z\, dz$$

EXAMPLE 10—

Find the approximate and total change for $u = x^2 y$, where we go from u(2,3) to u(1.97,3.02).

$\Delta x = dx = -.03$; $\Delta y = dy = .02$ $x = 2$; $y = 3$.

The actual change $\Delta u = u(1.97,3.02) - u(2,3) = (1.97)^2(3.02) - 2^2 3$.

If my calculator is OK, the answer is $11.720318 - 12 = -0.279682$.

$du = f_x\, dx + f_y\, df = 2xy\, dx + x^2\, dy = 2(2)(3)(-.03) + 2^2(.02) = -.36 + .08 = -0.28$, which is really a very good approximation, since (1.97,3.02) is very close to (2,3).

There are several places we can go now. Let's do the chain rule.

CHAIN RULE IN TWO (OR MORE) VARIABLES

Let $u = u(x,y)$ a function of two variables and $x = x(t)$ and $y = y(t)$. Then $du/dt = u_x\, dx/dt + u_y\, dy/dt$.

PROOF

$$\Delta u = u_x \Delta x + u_y \Delta y + \varepsilon_1 \Delta x + \varepsilon_2 \Delta y$$

$$\frac{\Delta u}{\Delta t} = u_x \frac{\Delta x}{\Delta t} + u_y \frac{\Delta y}{\Delta t} \varepsilon_1 \frac{\Delta x}{\Delta t} + \varepsilon_2 \frac{\Delta y}{\Delta t}$$

Divide by Δt.

$$du/dt = u_x\, dx/dt + u_y\, dy/dt$$

Limit $\Delta t \to 0$, $\varepsilon_1, \varepsilon_2$ also go to 0.

NOTE 1

This could extend to three or more variables.

NOTE 2

If $u = u(x,y)$ and $x = x(s,t)$ and $y = y(s,t)$, then

$$\frac{\partial u}{\partial s} = \frac{\partial u}{\partial x}\frac{\partial x}{\partial s} + \frac{\partial u}{\partial y}\frac{\partial y}{\partial s} \quad \text{and}$$

$$\frac{\partial u}{\partial t} = \frac{\partial u}{\partial x}\frac{\partial x}{\partial t} + \frac{\partial u}{\partial y}\frac{\partial y}{\partial t}$$

EXAMPLE 11—

Let $u = x^3y^4$, where $x = t^4$ and $y = t^{10}$.

$$du/dt = u_x\, dx/dt + u_y\, dy/dt$$
$$= (3x^2y^4)(4t^3) + (4x^3y^3)(10t^9)$$
$$= 3(t^4)^2(10t^{10})^4(4t^3) + 4(t^4)^3(t^{10})^3(10t^9) = 52t^{51}.$$

Direct substitution (not always easy) gives us a check.

$$u = x^3y^4 = (t^4)^3(t^{10})^4 = t^{52} \qquad du/dt = 52t^{51}$$

EXAMPLE 12—

$u = x^4 + y^5;\ x = s^2 + t^4;\ y = st^2$

$$\frac{\partial u}{\partial s} = \frac{\partial u}{\partial x}\frac{\partial x}{\partial s} + \frac{\partial u}{\partial y}\frac{\partial y}{\partial s}$$
$$= 4x^3(2s) + (5y^4)t^2$$

$$\frac{\partial u}{\partial t} = \frac{\partial u}{\partial x}\frac{\partial x}{\partial t} + \frac{\partial u}{\partial y}\frac{\partial y}{\partial t}$$
$$= (4x^3)(4t^3) + (5y^4)(2st)$$

Since insanity has run in my family for generations, I will put the second partials here. For variety, we will use r and s instead of t and s.

$$\frac{\partial u}{\partial r} = \frac{\partial u}{\partial x}\frac{\partial x}{\partial r} + \frac{\partial u}{\partial y}\frac{\partial y}{\partial r} \qquad \frac{\partial u}{\partial s} = \frac{\partial u}{\partial x}\frac{\partial x}{\partial s} + \frac{\partial u}{\partial y}\frac{\partial y}{\partial s}$$

$$\frac{\partial^2 u}{\partial r^2} = \frac{\partial}{\partial r}\left(\frac{\partial u}{\partial r}\right)$$
$$= \left[\frac{\partial}{\partial r}\left(\frac{\partial u}{\partial x}\right)\right]\frac{\partial x}{\partial r} + \frac{\partial u}{\partial x}\frac{\partial^2 x}{\partial r^2} + \left[\frac{\partial}{\partial r}\left(\frac{\partial u}{\partial y}\right)\right]\frac{\partial u}{\partial r} + \frac{\partial u}{\partial y}\frac{\partial^2 y}{\partial r^2}$$

That was the *product rule.* Buuut

$$\frac{\partial}{\partial r}\left[\frac{\partial u}{\partial x}\right] = \frac{\partial^2 u}{\partial x^2}\frac{\partial x}{\partial r} + \frac{\partial^2 u}{\partial y \partial x}\frac{\partial y}{\partial r}$$

$$\frac{\partial}{\partial r}\left[\frac{\partial u}{\partial y}\right] = \frac{\partial^2 u}{\partial x \partial y}\frac{\partial x}{\partial r} + \frac{\partial^2 u}{\partial y^2}\frac{\partial y}{\partial r}$$

This is the *chain rule.* Substituting the chain rule into the product rule, we get (after multiplying)

$$\frac{\partial^2 u}{\partial r^2} = \frac{\partial^2 u}{\partial x^2}\left(\frac{\partial x}{\partial r}\right)^2 + \partial \frac{\partial^2 u}{\partial y \partial x}\frac{\partial x}{\partial r}\frac{\partial y}{\partial r} + \frac{\partial^2 u}{\partial y^2}\left(\frac{\partial y}{\partial t}\right)^2$$

$$+ \frac{\partial u}{\partial x}\frac{\partial^2 x}{\partial r^2} + \frac{\partial u}{\partial y}\frac{\partial^2 y}{\partial r^2}$$

Similarly, (ha, ha, ha, ha)

$$\frac{\partial^2 u}{\partial s \partial r} = \frac{\partial^2 u \partial x \partial x}{\partial x^2 \partial r \partial s} + \frac{\partial^2 u \partial y \partial x}{\partial y \partial x \partial s \partial r} + \frac{\partial^2 u}{\partial x \partial y}\frac{\partial x}{\partial s}\frac{\partial y}{\partial r}$$

$$+ \frac{\partial^2 u}{\partial y^2}\frac{\partial y}{\partial s}\frac{\partial u}{\partial r} + \frac{\partial u}{\partial x}\frac{\partial^2 x}{\partial s \partial r} + \frac{\partial u}{\partial y}\frac{\partial^2 y}{\partial s \partial r}$$

To find $\partial^2 u/\partial r \partial s$ and $\partial^2 u/\partial s^2$, interchange r with s and s with r in the rules given.

Using this is bad, and simplifying is worse. This takes lots of practice. Most courses save this for advanced calculus, but not all.

In doing most of the examples, you probably found that $f_{xy} = f_{yx}$. Since many of the examples use polynomials, it is indeed true. However, there is a theorem which tells us exactly when $f_{xy} = f_{yx}$.

EQUALITY OF MIXED PARTIALS

Given $u = f(x,y)$ and f_{xy} exists at P. f_x, f_y, f_{yx} and f_{yx} continuous are defined in a neighborhood of P(x,y).

The conclusion is that, not only does f_{xy} exist, but $f_{xy} = f_{yx}$. The proof is long and is left to advanced calculus.

NOTE

This is certainly true about polynomials in more than one variable, since all partials exist, differential and continuous everywhere.

CHAPTER 4

DERIVATIVES IN MANY DIRECTIONS

DIRECTIONAL DERIVATIVES

I've searched through a number of books and found many, many notations for directional derivatives. I chose this one. I think it's the clearest.

Given

1. $z = f(x,y)$; (x_0,y_0) in Domain.
2. (x_0,y_0) a fixed point on L
3. (x,y) any arbitrary point on L
4. L in direction of unit vector $\langle a,b \rangle$
5. The partials $f_x(x_0,y_0)$, $f_y(x_0,y_0)$ we know about.

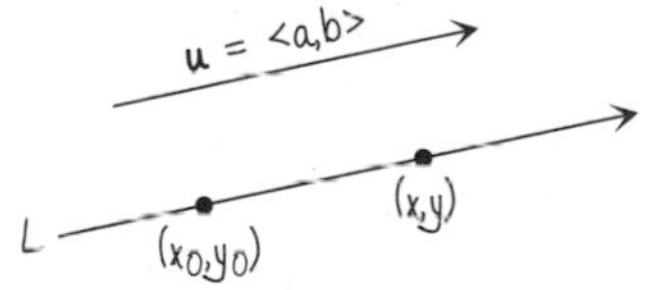

Then

1. parametric form of L is $x = x_0 + at$, $y = y_0 + bt$, $z = 0$.
2. The average change of f, with respect to t is given by the distance quotient

   $\dfrac{f(x_0 + at, y_0 + bt) - f(x_0,y_0)}{t}$. $\sqrt{(at)^2 + (bt)^2} =$

   $t\sqrt{a^2 + b^2} = t$ since $\langle a,b \rangle$ is a unit vector.

3. Taking the limit as t goes to 0, we get the instantaneous rate of change at (x_0,y_0) in the direction of **u**. Sooooo, let f be defined in a neighborhood around (x_0,y_0). Let $\mathbf{u} = \langle a,b \rangle$ be a unit vector. Then the *directional derivative* of f at (x_0,y_0) in the direction of **u** is

$$D_{\mathbf{u}}f(x_0,y_0) = \lim_{t\to 0} \frac{f(x_0 + at, y_0 + bt) - f(x_0,y_0)}{t}$$

if it exists.

NOTE 1
Note how similar all the derivative definitions are.

NOTE 2
If $\mathbf{u} = i$, then $a = 1$, $b = 0$, and $D_i = f_x$, the partial derivative. (Show!!!)

NOTE 3
If $\mathbf{u} = j$, then $D_j = f_y$.

A more usable form the directional derivative, given as a definition in some books and a theorem in others, is the following:

$$D_{\mathbf{u}}f(x_0,y_0) = af_x(x_0,y_0) + bf_y(x_0,y_0)$$

NOTE 4
We can extend these to three (or more) variables.

$\mathbf{u} = \langle a,b,c \rangle$

$$D_{\mathbf{u}}f(x_0,y_0,z_0) = \lim_{t\to 0} \frac{f(x_0 + at, y_0 + bt, z_0 + ct) - f(x_0,y_0,z_0)}{t}$$

Also

$$D_{\mathbf{u}}f = af_x + bf_y + cf_z$$

Now let's do some problems!!!!!!

EXAMPLE 1—

Find the directional derivative of $f(x,y) = x^2 + xy^2$ at (3,2) in the direction of $\mathbf{u} = \langle ½, \sqrt{3}/2 \rangle$.

$f_x = 2x + y^2$; $f_x(3,2) = 10$; $f_y = 2xy$; $f_y(3,2) = 12$

$D_\mathbf{u}f = af_x + bf_y = ½(10) + \sqrt{3/2}\,(12) = 5 + 6\sqrt{3}$

EXAMPLE 2—

Find $D_\mathbf{u}$ if $f(x,y) = e^{xy}$ at (7,6) in the direction of point (10,2).

$f_x = ye^{xy}$; $f_x(7,6) = 6e^{42}$; $f_y = xe^{xy}$; $f_y = 7e^{42}$

We must calculate **u**.

$\mathbf{v} = \langle 10 - 7, 2 - 6 \rangle = \langle 3,-4 \rangle$; $\mathbf{u} = \mathbf{v}/|\mathbf{v}| = \langle 3/5,-4/5 \rangle$

$D_\mathbf{u}f = af_x + bf_y = (3/5)6e^{42} + (-4/5)7e^{42} = -2e^{42}$

It is convenient to introduce a new term because it occurs so often: the *gradient* of f (written grad f, ∇f, or del f):

grad $f = \nabla f = f_x i + f_y j\ (+f_z k)$

IMPORTANT NOTE

$D_\mathbf{u}f = \nabla f \cdot \mathbf{u}$, the dot product of grad f and u!!!!

EXAMPLE 3—

$f(x,y) = y^2 + 2xy^3$. $(x_0,y_0) = (5,2)$. $\mathbf{u} = \langle .6,.8 \rangle$

A. Find the directional derivative.

B. Find the greatest rate of increase of f.

C. Find the greatest rate of decrease of f.

D. Find a direction orthogonal to del f.

A. $f_x = 2y^3 = 16$; $f_y = 2y + 6xy^2 = 124$; $\nabla f = 16i + 124j$.

$D_\mathbf{u}f = \nabla f \cdot \mathbf{u} = (16i + 124j) \cdot (.6i + .8j) = 108.8$.

B. $D_{\mathbf{u}}f = \nabla f \cdot \mathbf{u} = |\nabla f|\,|\mathbf{u}| \cos\theta$. Max. occurs when $\cos\theta = 1$.

$|\mathbf{u}|$ is always 1. Max is $|\nabla f|$.

$|\nabla f| = \sqrt{[124^2 + 16^2]} = 125.028$.

C. $-|\nabla f| = -125.028$ ($\cos\theta = -1$).

D. Orthogonal, perpendicular, no change—all mean that the dot product is 0. We did this before. $(16i + 124j) \cdot (ci + dj) = 0$. One such vector is $124i - 16j$ or $31i - 4j$.

The equation of a plane tangent to the surface $S = F(x,y,z) = 0$ is given by $\nabla F \cdot \langle x - x_0, y - y_0, z - z_0\rangle = 0$.

NOTE 1

A more usable form is $F_x(x_0,y_0,z_0)(x - x_0) + F_y(y - y_0) + F_z(z - z_0) = 0$.

NOTE 2

The proof is in many books. We already proved the theorem in the special case where $z = f(x,y)$.

Therefore, $F(x,y,z) = f(x,y) - z$. $F_x = f_x$, $F_y = f_y$, and $F_z = -1$. Therefore, at (x_0,y_0,z_0), the equation of tangent plane is

$$f_x(x - x_0) + f_y(y - y_0) - 1(z - z_0) = 0$$

orrrr

$$z - z_0 = f_x(x - x_0) + f_y(y - y_0)$$

EXAMPLE 4—

$z = x^2 + y^3$. Find the equation of the tangent plane at (1,2,9).

$f_x = 2x = 2$, $f_y = 3y^2 = 12$.

$z - z_0 = f_x(x - x_0) + f_y(y - y_0)$ $\quad z - 9 = 2(x - 1) + 12(y - 2)$

or $2x + 12y - z = 17$

EXAMPLE 5—

Find the point(s) on ellipsoid $x^2 + 4y^2 + 25z^2 = 405$ where the tangent plane is parallel to the plane $8x - 8y + 50z = 101$.

The problem is a little more involved. "Tangent planes" means the normal vectors of the plane are scalar multiples of each other. So $\nabla f = t\mathbf{N}_{plane}\langle 2x,8y,50z\rangle = t\langle 8,-8,50\rangle$. So $x = 4t$, $y = -t$, $z = t$.

In addition, x, y, and z must be on the ellipsoid. So $x^2 + 4y^2 + 25z^2 = (4t)^2 + 4(-t)^2 + 25t^2 = 45t^2 = 405$. $t = \pm 3$. So the points are ± 3 times $(4, -1,1)$. So the points are $(12,-3,3)$ and $(-12,3,-3)$.

IMPLICIT DIFFERENTIATION

Remember how much trouble we had differentiating implicitly? We can now do it in one step!!!!!

Suppose $F(x,y) = 0$. The trick is to let $x = t$. We will find dF/dt.

$$\frac{dF}{dt} = F_x(dx/dt) + F_y(dy/dt)$$

$F = 0$; $dF/dt = 0$. $dx/dt = dx/dx = 1$ and $dy/dt = dy/dx$!! Solving for dy/dx, we get $dy/dx = -F_x/F_y$.

EXAMPLE 6—

$x^2 + y^3 = x^5y^7 + 1$. Find dy/dx.

$F(x,y) = x^2 + y^3 - x^5y^7 - 1$

$F_x = 2x - 5x^4y^7 \qquad F_y = 3y^2 - 7x^5y^6$

$dy/dx = -F_x/F_y = -(2x - 5x^4y^7)/(3y^2 - 7x^5y^6)$

You might ask, "Why didn't we learn this earlier?" You might even be very angry. However, there are several reasons. First, there was great algebraic value in not knowing this. Second, the concept of a partial derivative (3D) is a jump in knowledge that few would be ready for. Math does not go from easy to hard; it goes from topic to topic and level to level.

This can be extended three or more unknowns.

$F(x,y,z) = 0 \qquad z_x = -F_x/F_z$ and $z_y = -F_y/F_z$

MAXIMUM AND MINIMUM IN TWO VARIABLES

We will do this topic very lightly. To do more advanced work and proofs, see some other elementary calc books and advanced calc books.

A *relative max* (*min*) is the highest (lowest) z value in a region. Here is a theorem to find and test for such points given point (a,b):

$f_x(a,b) = f_y(a,b) = 0$

assuming that fxx, fxy, and fyy are continuous in a neighborhood around (a,b).

Let $D = f_{xx}f_{yy}-f^2_{xy}$.

A. $D > 0$ and $f_{xx} > 0$, rel min

B. $D > 0$ and $f_{xx} < 0$, rel max

C. $D < 0$, saddle (like on a horse—a 3D extension of the inflection point, sort of)

D. $D = 0$??????????

NOTE

There are other max and min values. Sometimes, what it is can be determined by a picture if it can be sanely drawn.

Let's do two examples.

EXAMPLE 7—

$f(x,y) = x^6 + y^6 - 6xy$

$f_x = 6x^5 - 6y = 0 \quad y = x^5 \quad f_y = 6y^5 - 6x = 0$

$x = y^5 \quad x^{25} - x = 0$. SO $x = 1,-1,0$.

The critical points are (0,0), (1,1), (–1,–1).

(a,b)	$f_{xx}(a,b)$	$f_{yy}(a,b)$	$f_{xy}(a,b)$	$D = f_{xx}f_{yy} - f^2_{xy}$	
(0,0)	0	0	–6	–36	saddle
(1,1)	30	30	–6	864	rel min
(–1,–1)	30	30	–6	864	rel min

[$f_{xx} = 30x^4$ $f_{yy} = 30y^4$ $f_{xy} = -6$]

EXAMPLE 8—

Find the dimensions of the box from *Calc I* with a volume of 32 cubic inches and minimum surface area if the box has no top.

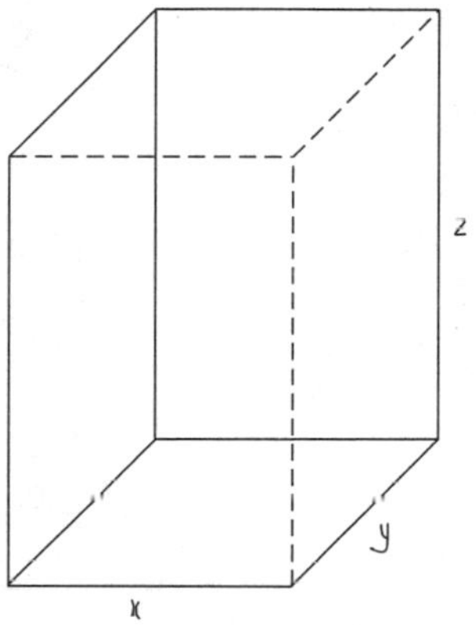

$V = xyz = 32$ $z = 32/xy$

surface area = xy + 2yz + 2xz

(bottom) (ends) (front, back)

$S = xy + 2y(32/xy) + 2x(32/xy)$

$= xy + 64/x + 64/y$

$S_x = y - 64/x^2 = 0$

$S_y = x - 64/y^2 = 0$

$xy^2 = 64$ and $x^2y = 64$. So $xy^2 = x^2y$. Since x or y can't be 0, then x = y and $x^3 = 64$. So x = 4 and y = 4.
$z = 32/xy = 2$.

Remember, when we did this problem in *Calc I,* we had to assume this was a square base. Now with more knowledge, we didn't have to assume as much. We will do this problem one more time a little later.

LA GRANGE MULTIPLIERS

This topic can be found in many courses and can be omitted from many courses. Let's do it in case you need it. Most of you will, eventually.

We wish to find the extreme values of f, say at the point (x_0,y_0), given a restrictive condition, call it g(x).

La Grange's theorem states that, if the grad g is not the zero vector at (x_0,y_0), there exists a number λ (lambda, a Greek letter) such that grad f = λ grad g. Translation into English: extreme values occur where their tangent planes at (x_0,y_0) are parallel; that is, their normals are proportional.

Let us give three examples.

EXAMPLE 9—

Find the maximum product of x and y on $x^2 + y^2 = 32$.

$f(x,y) = xy$ condition $g(x,y) = x^2 + y^2 - 32$

grad $f = \langle y,x \rangle$ grad $g = \langle 2x,2y \rangle$ $\nabla f = \lambda \nabla g$

$y = \lambda 2x$ $x = \lambda 2y$ $\lambda = y/2x$ $\lambda = x/2y$

Therefore

$y/2x = x/2y$ $2y^2 = 2x^2$

So

$x^2 = y^2$

Substituting in the circle, $2x^2 = 32$. $x^2 = 16$; $x = \pm 4$; $y = \pm 4$. There are four possible points: (4,4), (4,–4), (–4,4), and (–4,–4). f(4,4) = 16, f(–4,–4) = 16, f(4,–4) = –16, and f(–4,4) = –16.

Therefore, (4,4) and (–4,–4) are maxes.

EXAMPLE 10—

Find the minimum distance from $x + 2y + 3z = 27$ to (2,3,4).

To avoid square roots, we will use the same trick as we used in *Calc I:* we will minimize the square of the distance formula.

$f(x,y,z) = (x-2)^2 + (y-3)^2 + (z-4)^2 \qquad \nabla f = \langle 2(x-2), 2(y-3), 2(z-4)\rangle$

$g(x,y,z) = x + 2y + 3z - 27 \qquad \nabla g = \langle 1,2,3\rangle.$

$2(x-2) = \lambda 1 \qquad x = (4+\lambda)/2 \qquad \nabla f = \lambda \nabla g$

$2(y-3) = \lambda 2 \qquad y = (2\lambda + 6)/2$

$2(z-4) = \lambda 3 \qquad z = (3\lambda + 8)/2$

Substituting into $x + 2y + 3z = 27$, we get $(4+\lambda)/2 + 2(2\lambda+6)/2 + 3(3\lambda+8)/2 = 27$. We get $\lambda = 1$. $x = (4+1)/2 = 5/2$. $y = (2(1)+6)/2 = 4$. $z = (3(1)+8)/2 = 11/2$.

The point is $(5/2, 4, 11/2)$.

Lastly, let's do the same problem we've done twice before; this time we use La Grange multipliers. Yes, find the minimum surface area of a topless box with a volume of 32 cubic light years.

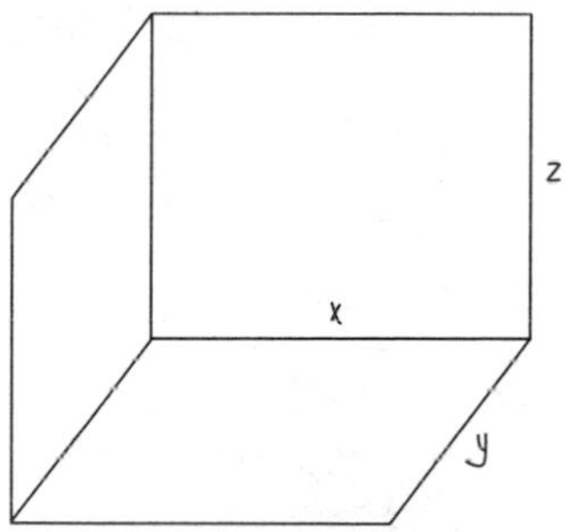

Let the surface area $f(x,y,z) = xy + 2xz + 2yz$, subject to the condition $g(x,y,z) = xyz - 32$.

$\nabla f = \langle y + 2z, x + 2z, 2y + 2z\rangle$

$\nabla g = \langle yz, xz, xy\rangle \qquad \nabla f = \lambda \nabla g$

A. $y + 2z = \lambda yz$

B. $x + 2z = \lambda xz$

C. $2y + 2x = \lambda xy$

D. $xyz = 12$

A – B. $y - x = \lambda z(y - x)$. Two possibilities: (1) $\lambda z = 1$ or (2) $x = y$. Possibility (1) can't happen since Eq. A would result in $z = 0$, which is impossible. Therefore, $x = y$. Substitute $x = y$ in C. We get $\lambda = 4/x$. Substitute this in B. We get $x + 2z = (4/x)xz$ or $x = 2z$ or $z = x/2$.

Therefore, $xyz = x(x)(x/2) = 32$. $x^3 = 64$. $x = 4$. $y = x = 4$. $z = 4/2 = 2$. And again our box is 4 by 4 by 2 (light years).

The one thing that makes La Grange multipliers difficult is the algebra. The setups are very similar, but unfortunately no one algebraic method works. Each problem requires skill and ingenuity of its own. These three problems will give you a sampling. But at this point in your mathematics career, you should be up to the challenge.

CHAPTER 5

DOUBLE INTEGRALS: x-y AND POLAR

MULTIPLE INTEGRALS

The next level of integration, areas, and volumes involves two or three integrations in the same problem. My belief is that, if you understand double integration with x-y coordinates, you will understand everything. The pressure is on me to write the very best I can here.

But first, the motivation. If you go back to my *Calc I Helper,* you will see that the motivation for the single integral is the area under the curve. We will do the same for the double integral, but not in as much detail.

Given a region with a base in the x-y plane and a roof $z = f(x,y)$, R is the projection of the roof into the x-y plane. Make a grid in the x-y plane; usually squares are easiest for computation. The grid can be completely inside R, overlap the border of R in places, or totally overlap the border; usually completely inside is easiest for computation. At some point in ΔA_k, find the point (u_k,v_k). The height of this tall, thin box is $z = f(u_k,v_k)$. The volume $\Delta V_k = f(u_k,v_k) \cdot \Delta A_k$. Add them all up. We get $\sum_{k=1}^{n} f(u_k,v_k)\Delta A_k$.

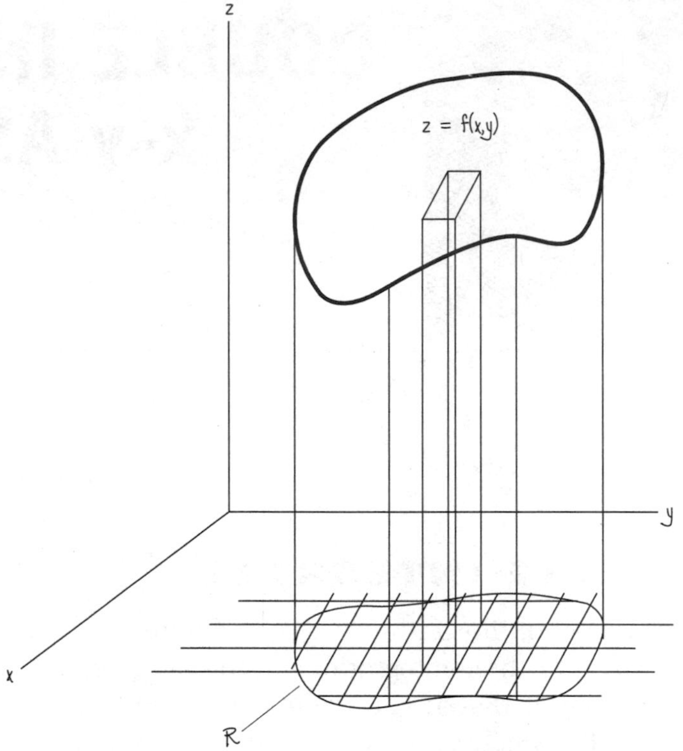

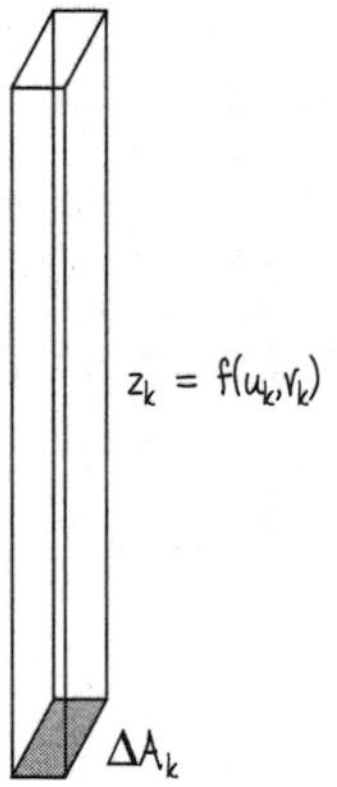

In the limit as ΔA_k, go to 0, letting both the length and the width of ΔA_k (if it's a square) or all measurements (if it's something else) go to 0. There will be more and more "squares," which will come closer and closer to the entire region R. In the limit, as n goes to infinity, we get the volume, which is

$$\iint_R f(x,y)\, dA$$

We now define the double integral in exactly the same way, forgetting the figure.

However, just like in the case of the single integral, this procedure is virtually impossible to use, since it is incredibly long and complicated. Again, fortunately, there is a way out.

Integrated Integrals: x-y Coordinates

We would like to examine $\int_a^b f(x,y)\,dx$. Look at the figure. We are adding up little Δx's. What do you see? y does *not* change. So, in integrating with respect to x, we hold y constant. Let a = 2 and b = 5.

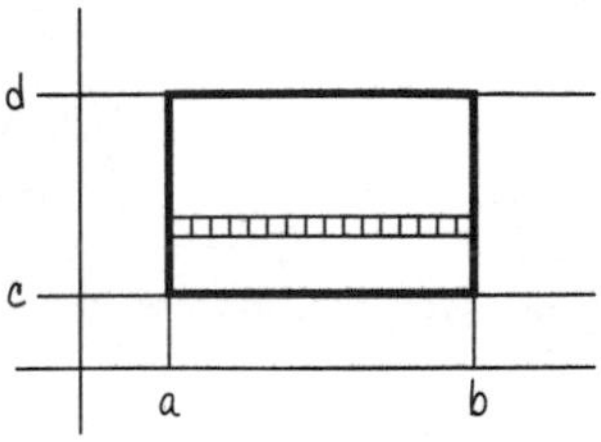

EXAMPLE 1—

$$\int_{x=2}^{5} (3x^2 + 2y + 6xy)\,dx$$

$$= x^3 + 2xy + 3x^2y \Big|_{x=2}^{x=5}$$

Remember, in the second and third term, y is a constant.

$$= (5^3 + 2(5)y + 3(5^2)y) - (2^3 + 2(2)y + 3(2)^2y)$$

$$= 117 + 69y$$

Look at the second figure. Notice x does not change. Thereforeeeee. . . .

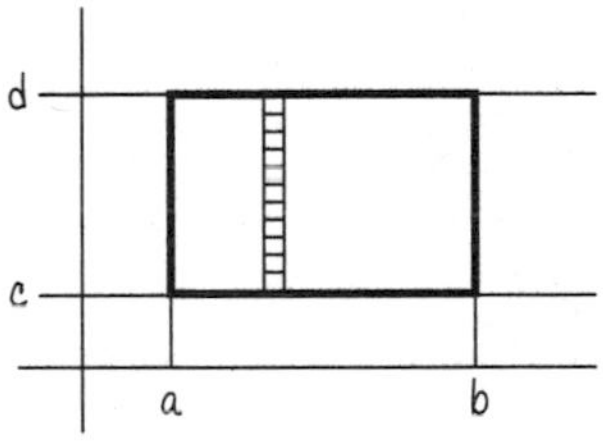

EXAMPLE 2—

Let c = 0 and d = 4.

$$\int_{y=0}^{4} (3x^2 + 2y + 6xy)\,dy$$

$$= 3x^2y + y^2 + 3xy^2 \Big|_{y=0}^{4}$$

Remember in terms 1 and 3, x is a constant.

$$= 12x^2 + 48x + 16$$

OK so far? Let's keep going.

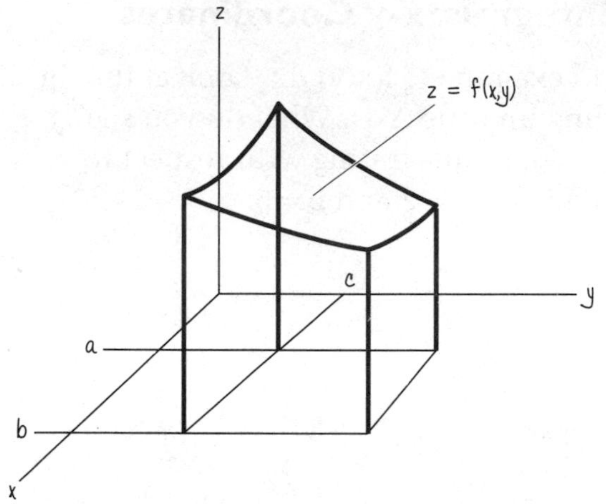

If we do the integral $\int_a^b f(x,y)\,dx$, what we get is an area that is a function of y only, call it A(y).

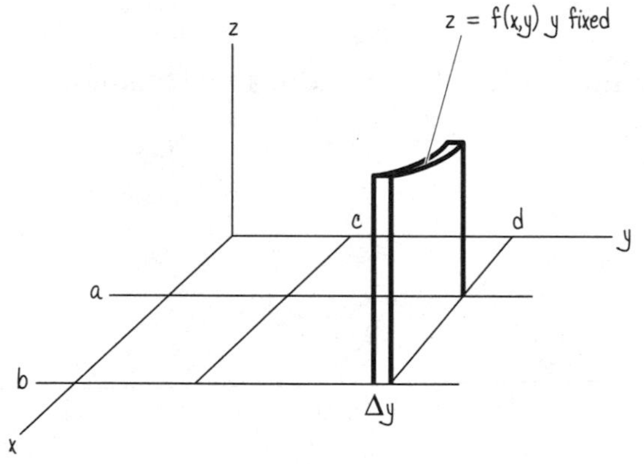

If we then take A(y) and multiply it by Δy, we get the volume of that thin strip. If we add all those strips and take the appropriate limits, we will get the volume of the whole region. Symbolically,

$\int_{y=c}^{d} A(y)\,dy = V.$ All together

$$\int_{y=c}^{d}\int_{x=a}^{b} f(x,y)\,dx\,dy = V$$

NOTE
This is very similar to finding volumes by sections. In fact, in the second part, after we get A(y), it is.
Similarly, with $\int_c^d f(x,y)\,dy$, we will get an area A(x). Multiply it by Δx, and again follow the procedure before and we get the volume.

$$\int_{x=a}^{b} A(x)\,dx = V$$

$$V = \int_a^b \int_c^d f(x,y)\,dy\,dx$$

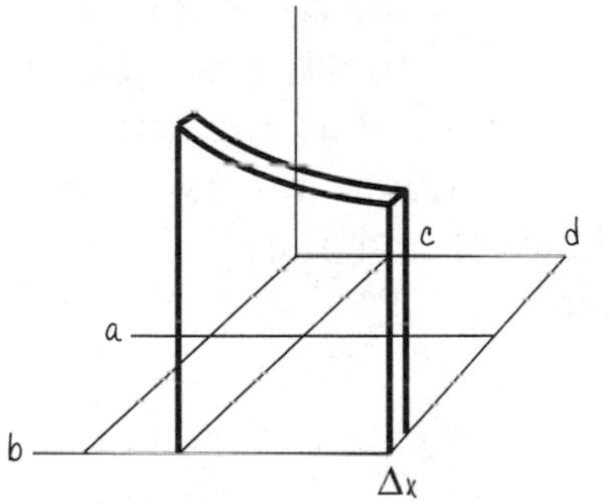

Since it is the name volume, all three double integrals must be the same.

$$\iint_R f(x,y)\,dA = \int_c^d \int_a^b f(x,y)\,dx\,dy = \int_a^b \int_c^d f(x,y)\,dy\,dx$$

Let us finish Example 1.

EXAMPLE 1 (CONTINUED)—

$$\int_{y=0}^{4} \int_{x=2}^{5} (3x^2 + 2y + 6xy)\,dx\,dy$$

$$= \int_{y=0}^{4} (117 + 69y)\,dy = 117y + \frac{69y^2}{2}\Big[_0^4$$

$$= 468 + 552 = 1020$$

EXAMPLE 2 (CONTINUED)—

$$\int_{x=2}^{5}\int_{y=0}^{4} (3x^2 + 2y + 6xy)\,dy\,dx$$

$$= \int_{x=2}^{5} (12x^2 + 48x + 16)\,dx = 4x^3 + 24x^2 + 16x \Big[_{2}^{5}$$

$$= 1180 - 160 = 1020$$

The two answers agree as they must. The volume is 1020.

OK so far? If not, reread several times.

Let's go on.

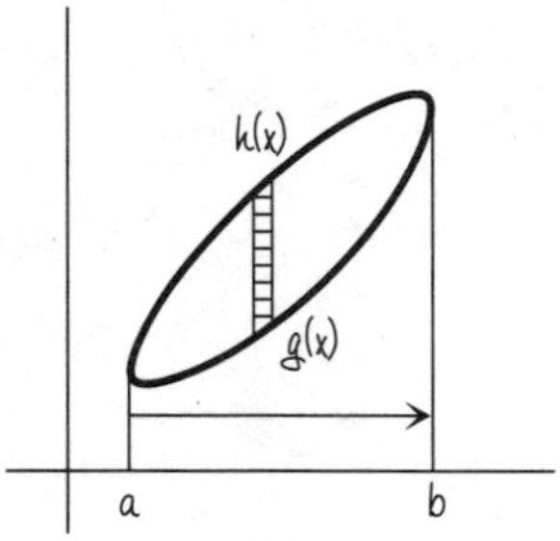

What if the region is not a rectangle? We will do the double integral in a similar manner to when we found the area in *Calc I.* Again, fix the x, add up the little changes in y from g(x) to h(x), and then integrate from a to b to get the volume. You must imagine the height z = f(x,y), which is not drawn.

$$\int_{x=a}^{b}\left[\int_{y=g(x)}^{h(x)} f(x,y)\,dy\right]dx$$

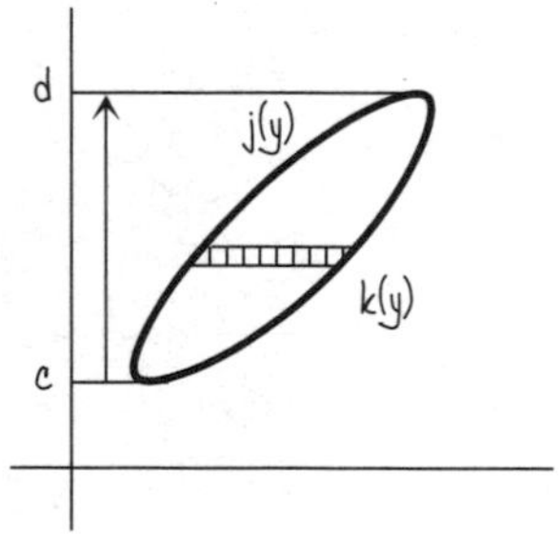

Similarly, we get the integral

$$\int_{y=c}^{d}\left[\int_{x=j(y)}^{k(y)} f(x,y)\,dx\right]dy$$

Let's do a problem two different ways. Let z = f(x,y) be the height (z = 0 is the base), where z = 6x + 12y. The region in the x-y plane is bounded by y = 2x and $y = x^2$. We will integrate first with regard to y.

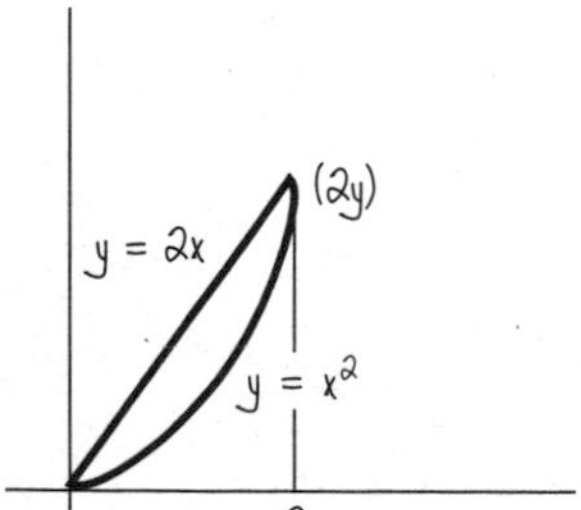

$$\int_{x=0}^{2}\int_{y=x^2}^{2x} (6x + 12y)\,dy\,dx = 6xy + y^2 \Big[_{y=x^2}^{y=2x}$$

$$= 6x(2x) + (2x)^2 - (6x(x^2) + (6x^2)^2)$$

$$\int_{x=0}^{2} (36x^2 - 6x^3 - 6x^4)\,dx = 12x^3 - \frac{3}{2}x^4 - \frac{6}{5}x^5 \Big[_{0}^{2}$$

$$= 96 - 24 - 38.4 = 33.6$$

Similarly, using the next figure, we get. . . .

$$\int_{y=0}^{4}\int_{x=y/2}^{\sqrt{y}} (6x + 12y)\,dx\,dy$$

$$= \int_{y=0}^{4} 3x^2 + 12xy \Big[_{x=y/2}^{\sqrt{y}}\ dy$$

$$= \int_{y=0}^{4} 3(\sqrt{y})^2 + 12(\sqrt{y})y - \left(3\left(\frac{y}{2}\right)^2 + 12\left(\frac{y}{2}\right)y\right) dy$$

$$= \int_{y=0}^{4} \left(3y + 12y^{3/2} - \frac{27y^2}{2}\right) dy$$

$$= \frac{3}{2}y^2 + \frac{24}{5}y^{5/2} - \frac{9y^3}{2}\Big[_{0}^{4}$$

$$= 24 + 153.6 - 144 = 33.6$$

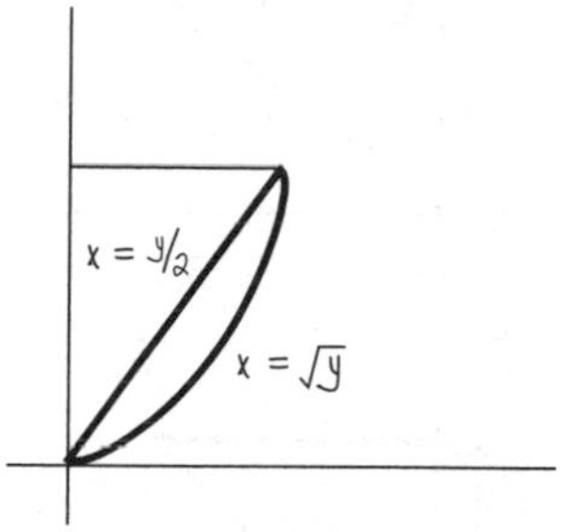

and the two answers agree!!!!

NOTE 1

If f(x,y) = 1, we found a way to find the area!!!!!!

NOTE 2

The reason this is so difficult is that x and y are "equals." That is, we can integrate first with respect to x or with respect to y, depending on the problem. Which should we do first? Let us do two problems, one all the way out, to answer this question.

EXAMPLE 4—

Let us say we have some z = f(x,y). Suppose the region is the one in the figure. In most cases, integrating with respect to y first is easier, since if we integrated with respect to x first, we would have to split the region.

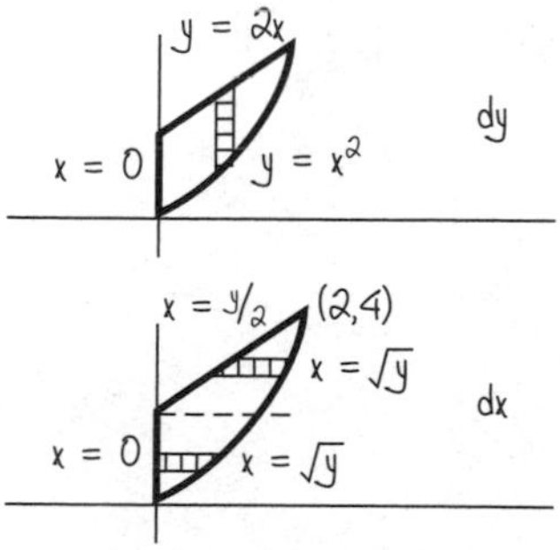

EXAMPLE 5—

Consider the double integral

$$\int_0^2 \int_x^2 6x\, e^{y^3}\, dy\, dx$$

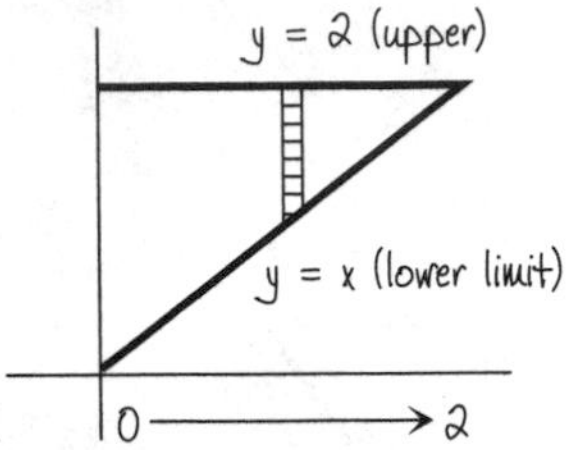

There is a very good reason why we don't integrate with respect to y first: we can't!!!!! It is not an integration we are capable of doing. So we want to reverse the limits. First, draw a picture of the region you are integrating over, the original way. Next, redraw, changing the limits. Finally, do the integration.

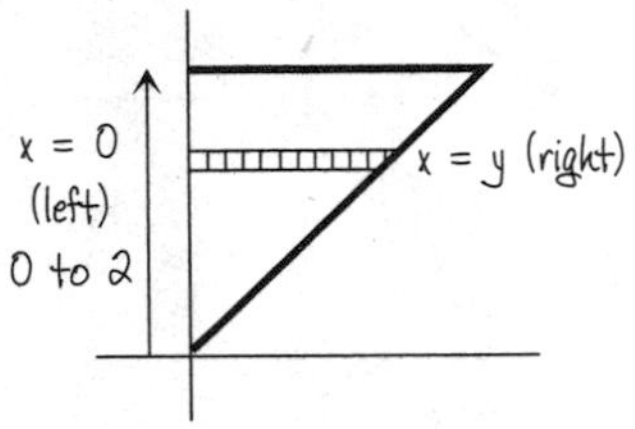

$$\int_{x=0}^{2} \int_{y=x}^{y=2} 6x\, e^{y^3}\, dy\, dx$$

$$= \int_{y=0}^{2} \int_{x=0}^{x=y} 6x\, e^{y^3}\, dx\, dy$$

$$= \int_{y=0}^{2} 3x^2\, e^{y^3} \Big|_{x=0}^{x=y} dy$$

$$= \int_{y=0}^{2} 3y^2\, e^{y^3}\, dy = e^{y^3} \Big|_0^2 = e^8 - 1 \qquad u = y^3 \qquad du = 3y^2 dy$$

Well, I hope you've got it. If you do, the rest of this chapter should not be too bad. The next double integral we shall do is that of polar coordinates. This should be easier because you will always integrate with respect to r first and then with respect to θ, the angle, last.

Polar Coordinates

This topic is a natural extension of polar coordinates from *Calc II.* Just like for x-y coordinates, we make a grid. Each unit of area is approximately a rectangle. The base is $r\Delta\theta$, and the height is Δr. $\Delta A_k = r_k \Delta r_k \Delta\theta_k$. Taking the proper sums and limits, we get . . .

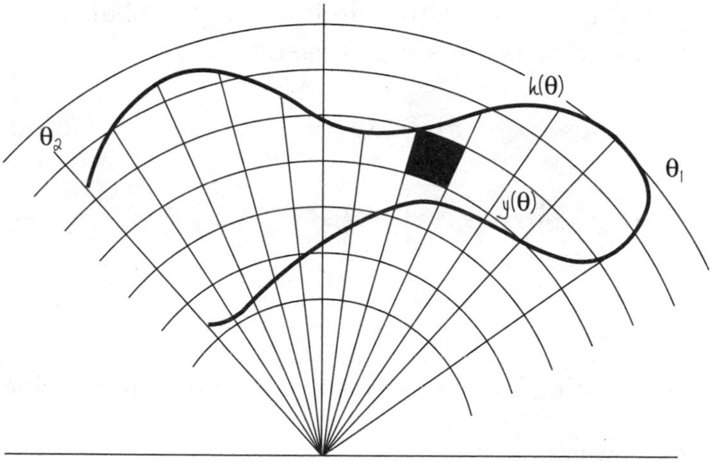

$$\int_{\theta=\theta_1}^{\theta 2} \int_{r=g(\theta)}^{h(\theta)} f(r,\theta)\, r\, dr\, d\theta$$

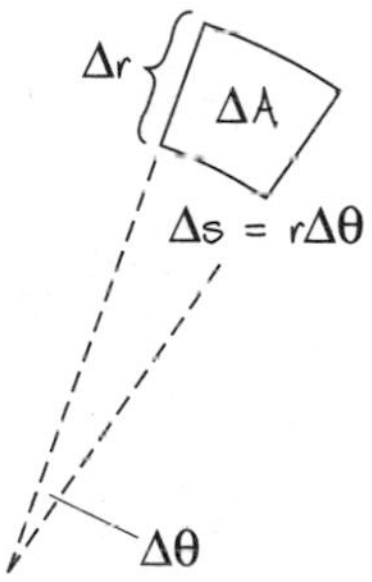

NOTE 1

$f(x,y)\,dx\,dy$ (or $dy\,dx$) is replaced by $f(r,\theta)\,r\,dr\,d\theta$.

NOTE 2

Just like in *Calc II*, we integrate outside (top limit) minus inside (bottom limit) with respect to r, and the angles go counterclockwise, with the larger angle on top and the smaller angle on bottom.

NOTE 3

If $f(r,\theta) = 1$, we get the area!!!!

EXAMPLE 6—

Find the integral of $\cos(x^2 + y^2)$ over the circle $x^2 + y^2 = 4$ in the first quadrant. If you set up the integral in x-y coordinates, we cannot do it, but if you change to polar, the work is done rather easily.

$x^2 + y^2 = r^2$

$$\int_{y=0}^{2}\int_{x=0}^{\sqrt{4-y^2}} \cos(x^2+y^2)\,dx\,dy = \int_{\theta=0}^{\pi/2}\int_{r=0}^{2} r\cos(r^2)\,dr\,d\theta$$

$dx\,dy = r\,dr\,d\theta$; 1st quad = 0 to $\pi/2$, no inside r, so r = 0.

$$= \int_{\theta=0}^{\pi/2} \frac{1}{2}\sin r^2 \Big|_{r=0}^{r=2} d\theta = \int_{0}^{\pi/2} \frac{1}{2}\sin 4\,d\theta = \frac{\pi}{4}\sin 4$$

NOTE

The integrating factor of r makes this integration doable.

EXAMPLE 7—

$$\int_{0}^{\pi/2}\int_{r=\cos\theta}^{1} 5r^4 \sin\theta\,dr\,d\theta$$

We are first going to integrate this dr—r changes, the angle stays the same. Then we will integrate dθ.

$$\int_{0}^{\pi/2}\int_{r=\cos\theta}^{1} 5r^4 \sin\theta\,dr\,d\theta$$

$$= \int_{0}^{\pi/2} r^5 \sin\theta \Big|_{r=\cos\theta}^{r=1} d\theta$$

$$= \int_{0}^{\pi/2} (\sin\theta - \cos^5\theta\sin\theta)\,d\theta = -\cos\theta + \frac{\cos^6\theta}{6}\Big|_{0}^{\pi/2} = \frac{5}{6}$$

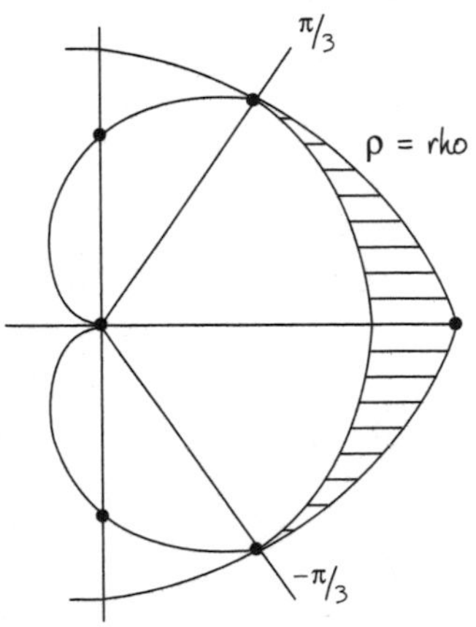

EXAMPLE 8—

Find the weight of the region inside r = 2 + 2 cos θ and outside r = 3 if the weight density ρ = 3r pounds per square foot.

1. Many density problems involve volume; that is, pounds per cubic feet.
2. This is a good time to throw in some review of Calc II.

3. Many of Calc III integrals have Calc I or simple Calc II integrals, but that doesn't mean you can forget the long Calc II integrals.

Now let's do the problem.

1. $r = 2 + 2\cos\theta$ is a cardioid. If it's inside the cardioid, the cardioid is the outside curve. $r = 3$, circle, is the inside curve, the lower limit.

2. To find the angles, set $r = 3 = 2 + 2\cos\theta$. The limits are $\pm\pi/3$.

$$\int_{\theta=-\pi/3}^{\pi/3}\int_{r=3}^{r=2+2\cos\theta} 3r \cdot r\,dr\,d\theta$$

3r = density; r dr dθ = integrating factor in polar

$$= \int_{-\pi/3}^{\pi/3} r^3 \Big|_{r=3}^{2+2\cos\theta} d\theta = \int_{-\pi/3}^{\pi/3} (2 + 2\cos\theta)^3 - 3^3\,d\theta$$

$$= \int_{-\pi/3}^{\pi/3} (24\cos\theta + 24\cos^2\theta + 8\cos^3\theta - 19)\,d\theta$$

$$= \int_{-\pi/3}^{\pi/3} (24\cos\theta + 24\left[\frac{1+\cos 2\theta}{2}\right]$$

$$+ 8\cos\theta\,[1 - \sin^2\theta] - 19)\,d\theta$$

$$= \int_{-\pi/3}^{\pi/3} (32\cos\theta + 12\cos 2\theta - 8\cos\theta\sin^2\theta - 7)\,d\theta$$

$$= 32\sin\theta + 6\sin^2\theta - \frac{8}{3}\sin^3\theta - 7\theta\Big]_{-\pi/3}^{\pi/3}$$

$$= 36\sqrt{3} - \frac{14\pi}{3}\text{ lbs}$$

Whew!

CHAPTER 6

TRIPLE INTEGRALS: x-y-z, CYLINDRICAL, SPHERICAL

Let's go to the triple integral. We extend the concept of double integral. Instead of giving the height, we have to calculate the height by taking little Δz's from the "floor" to the "ceiling" and then proceeding as before.

$$\int_{x=\text{back}}^{\text{front}} \int_{y=\text{left}}^{\text{right}} \int_{z=\text{floor}}^{z=\text{ceiling}} f(x,y,z)\, dz\, dy\, dx$$

$$\int_{x=a}^{b} \int_{y=k(x)}^{y=m(x)} \int_{z=g(x,y)}^{h(x,y)} f(x,y,z)\, dz\, dy\, dx$$

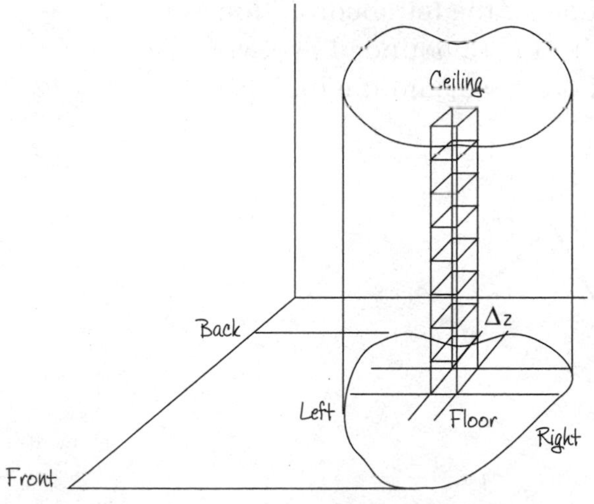

LOTS OF NOTES

1. We first integrate dz—z changes, leaving x and y fixed.
2. We then integrate dy, leaving x fixed (just like in a double integral) over the region where the z's meet. We set the ceiling equal to the floor.
3. We then integrate dx.
4. We could do dx first and then dy.
5. In triple integrals, we could do any of the variables first. For example, we could do dy dz dx—dy (holding x and z fixed), setting y's equal to each other, dz (holding x fixed), and finally dx.
6. In most modern books, dz is done first, since the figures are easiest to draw.
7. Fortunately, the regions in triple integrals are not too complicated. (Otherwise, we might never finish them.) Most can be done without drawing them (again, very fortunately).

Let's do an example.

EXAMPLE 1—

Find the volume of the tetrahedron (four-sided pyramid) $2x + 3y + 4z = 12$ bounded by x-y, x-z, and y-z planes. The floor-$z = 0$, and the roof $2x + 3y + 4z = 12$.

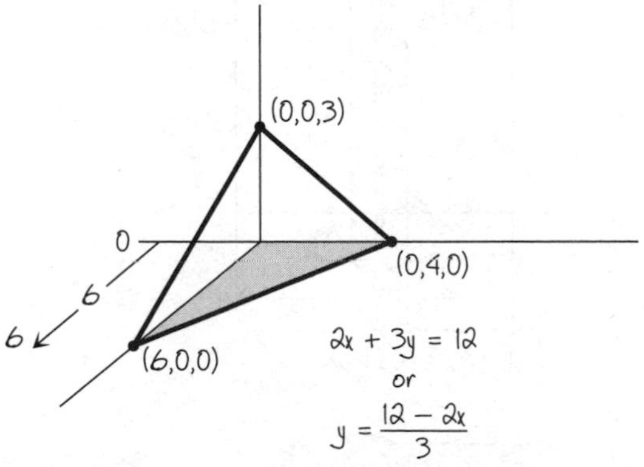

Setting them equal (in this case, substituting z = 0), we get the region in the second integral: y = 0 on the left, and y = (12 – 2x)/3 on the right. In the last integral (going right to left), x goes from 0 to 6.

$$\int_{x=0}^{6}\int_{y=0}^{(12-2x)/3}\int_{z=0}^{(12-2x-3y)/4} dz\,dy\,dx$$

$$= \int_{0}^{6}\int_{y=0}^{(12-2x)/3} z\Big[_{z=0}^{(12-2x-3y)/4} dy\,dx$$

$$= \int_{0}^{6}\int_{0}^{(12-2x)/3} \left(3 - \frac{x}{2} - \frac{3y}{4}\right) dy\,dx$$

$$= \int_{0}^{6} 3y - \frac{xy}{2} - \frac{3y^2}{8}\Big[_{y=0}^{y=(12-2x)/3}$$

$$= \int_{0}^{6} 3\left[\frac{12-2x}{3}\right] - \frac{x}{2}\left[\frac{12-2x}{3}\right] - \frac{3}{8}\left[\frac{12-2x}{3}\right]^2 dx$$

$$= \int_{0}^{6}\left(6 - 2x - \frac{x^2}{6}\right) dx = 6x - x^2 - \frac{x^3}{3}\Big[_{0}^{6} = 12$$

NOTE

In general, the volume of tetrahedron x/a + y/b + z/c = 1 in the first octant is V = 1/6 abc.

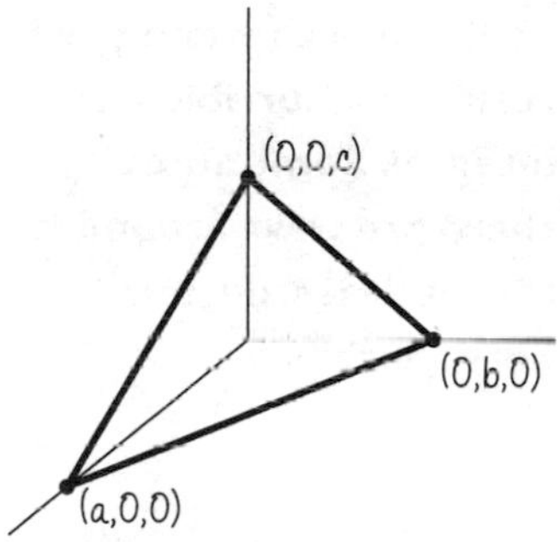

You may have seen this formula before. In the previous example, a = 6, b = 4, and c = 3. V = (1/6)(6)(4)(3) = 12. The actual verification isn't so nice, is it???!!!!

NOTE 1 (BEFORE THIS EXAMPLE)

In a triple integral, if f(x,y,z) = 1, we get the volume, as we have done above.

NOTE 2

f(x,y,z) could be a density, a distance to give us the torque, and many other uses.

Let us do one more example.

EXAMPLE 2—

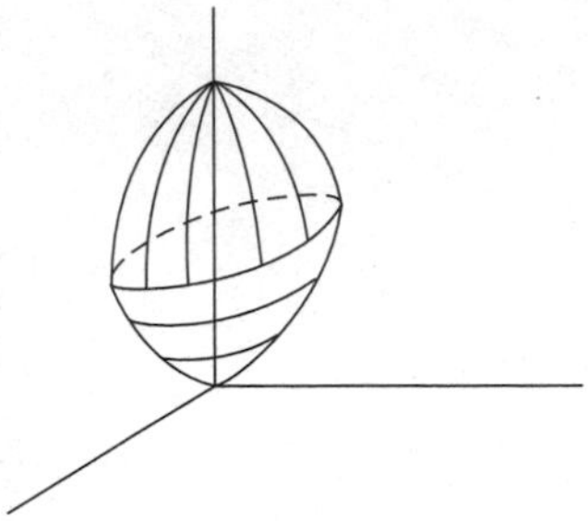

Find the volume between $z = x^2 + y^2$ and $z = 8 - x^2 - y^2$.

If we draw the figure, we find the roof is $z = 8 - x^2 - y^2$ and the floor is $z = x^2 + y^2$. However, many times we will not be able to draw. If the drawings do not present any unusual aspects, there is another way. If we let $x = 0$ and $y = 0$, in one case we get the point (0,0,0). In the other case, we get (0,0,8). Since (0,0,8) is on $y = 8 - x^2 - y^2$ and is above (0,0,0), $z = 8 - x^2 - y^2$ is the ceiling.

To find the area, we integrate (the double integral part), we set the z's equal: $x^2 + y^2 = 8 - x^2 - y^2$ or $x^2 + y^2 = 4$, then circle (center $x = 0$, $y = 0$; radius 2). The integral isssssssss.......

$$\int_{y=-2}^{2} \int_{x=-\sqrt{4-y^2}}^{x=\sqrt{4-y^2}} \int_{z=x^2+y^2 \text{ (floor)}}^{z=8-x^2-y^2 \text{ (roof)}} dz\, dx\, dy$$

$$= \int_{y=-2}^{2} \int_{x=-\sqrt{4-y^2}}^{+\sqrt{4-y^2}} z \Big[_{x^2+y^2}^{8-x^2-y^2} dx\, dy$$

To finish with x and y will result in a horrible last integral. Sometimes there are ways around it. (Sometimes, no, but here, yes!)

$$= \int_{y=-2}^{2} \int_{x=-\sqrt{4-y^2}}^{+\sqrt{4-y^2}} (8 - 2x^2 - 2y^2)\, dx\, dy$$

We can change to polar coordinates!!

$-2x^2 - 2y^2 = -2(x^2 + y^2) = -2r^2$; r dr dθ = integrating factor.

$$= \int_{\theta=0}^{2\pi} \int_{r=0}^{2} (8 - 2r^2)\, r\, dr\, d\theta$$

Easy, huh?!

$$= \int_{\theta=0}^{2\pi} 4r^2 - \frac{r^4}{2} \Big[_0^2 d\theta = \int_0^{2\pi} 8\, d\theta = 16\pi$$

which naturally brings us to the next topic..........

CYLINDRICAL COORDINATES

What we have just used is *cylindrical coordinates;* that is, $x = r \cos \theta$, $y = r \sin \theta$ and $z = z$, which is polar coordinates with $z = z$.

In the figures are the three constants in cylindrical coordinates: $r = k$ (a konstant—remember math people can't spell) is a cylinder. ($r = k$ in 2D is a circle.)

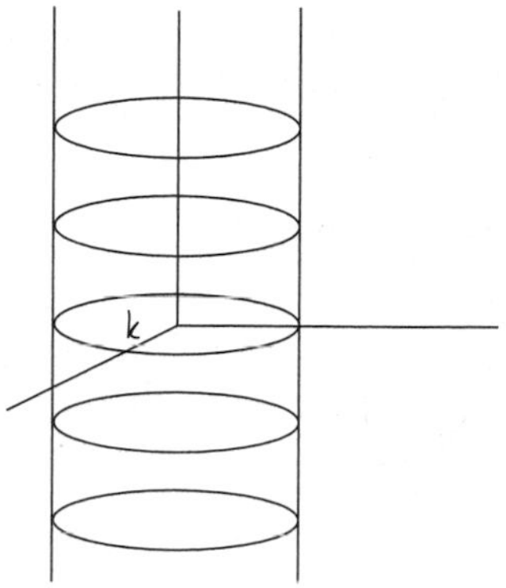

$\theta = k$ is a plane (in 2D it is a line).

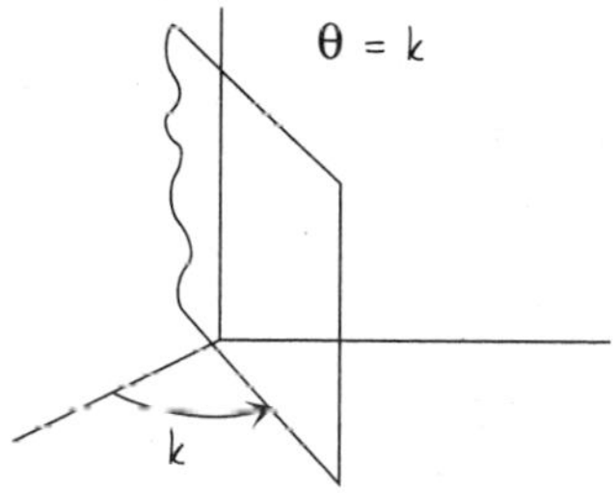

$z = k$ is also a plane. All three are shown, as if you needed me to tell you.

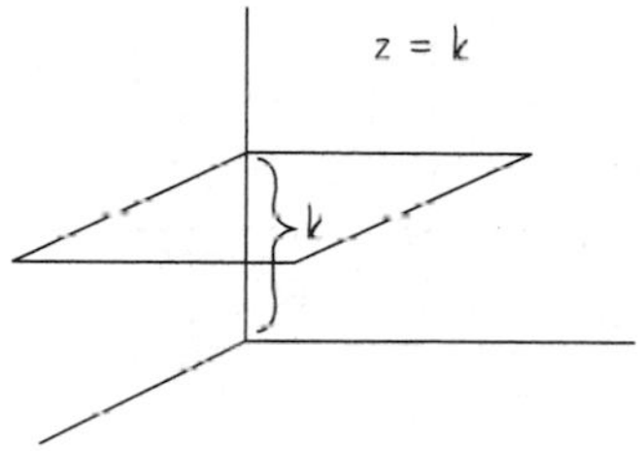

The next figure is a little unit of volume.

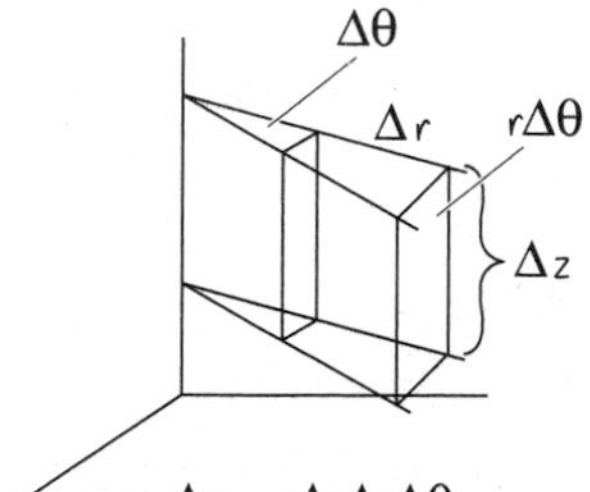

$f(x,y,z)\,dz\,dy\,dx$ becomes $f(r,\theta,z)\,r\,dz\,dr\,d\theta$, just like in the last example. Let's do one more.

EXAMPLE 3—

Find the volume of $z = 4 + x^2 + y^2$ inside the cylinder $x^2 + y^2 = 4$ in the first octant (x, y, and z are all ≥ 0).

$$\int_{y=0}^{2}\int_{x=0}^{\sqrt{4-y^2}}\int_{0}^{4+x^2+y^2} dz\,dx\,dy$$

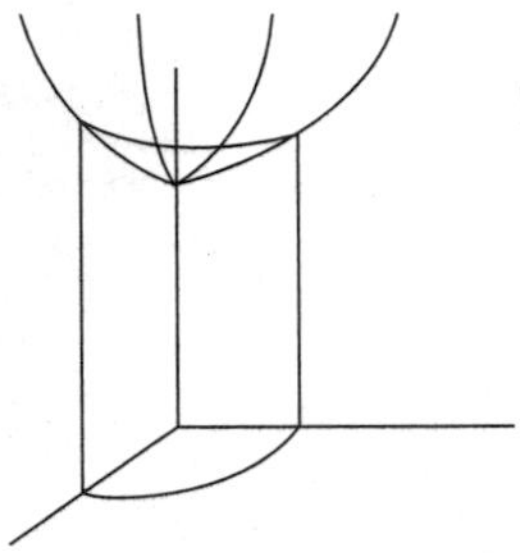

$$= \int_{\theta=0}^{\pi/2}\int_{r=0}^{2}\int_{z=0}^{4+r^2} r\,dz\,dr\,d\theta$$

$$= \int_{\theta=0}^{\pi/2}\int_{r=0}^{2} rz\Big[_{z=0}^{4+r^2} dr\,d\theta = \int_{0}^{2\pi}\int_{0}^{2} (4r + r^3)\,dr\,d\theta$$

$$= \int_{0}^{\pi/2} 2r^2 + \frac{r^4}{4}\Big[_{r=0}^{2} d\theta = \int_{0}^{\pi/2} 12\,d\theta = 6\pi$$

SPHERICAL COORDINATES

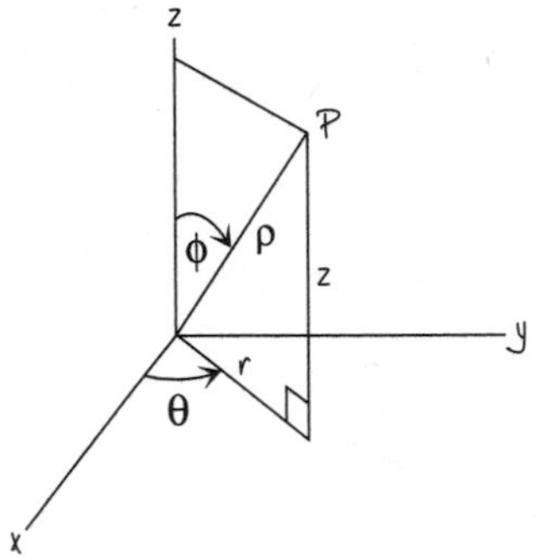

This topic is perhaps my least favorite, not because of the problems, but because of the derivations and the pictures you need to draw. (Although I am reasonably pleased with my original pictures, it took a long time to draw them so that the publisher's artists knew what I wanted.)

Given $f(\rho,\phi,\theta) = f(\text{rho,phi,theta})$

ρ is the distance from the origin (0,0,0) to the point (x,y,z), or $(x^2 + y^2 + z^2)^{1/2}$.

θ is the same as in polar coordinates—positive in a counterclockwise direction from zero to 2π.

ϕ is the angle measured in a positive sense from the positive z axis down and goes from 0 to π.

From the first figure on page 72, we see that $z = \rho \cos \phi$. The projection of ρ into the x-y plane, r, like in polar coordinates, is $r = \rho \sin \phi$.

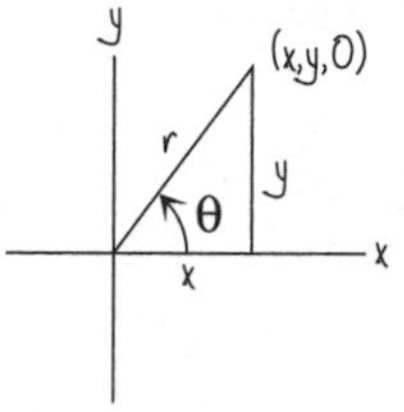

Looking at the second figure, we get $x = r \cos \theta$ and $y = r \sin \theta$. Substituting $\rho \sin \phi$ for r, we get $x = \rho \sin \phi \cos \theta$, $y = \rho \sin \phi \sin \theta$, and, of course, $z = \rho \cos \phi$.

In the next three figures, we see what each of the variables are as konstants, errr . . . *constants.* ρ, a constant, is a sphere, radius k about the origin. Φ, a constant, is a cone of angle k from the positive z axis. θ, as it is in polar coordinates as a constant, is a plane through the z axis at an angle θ away from the positive x axis.

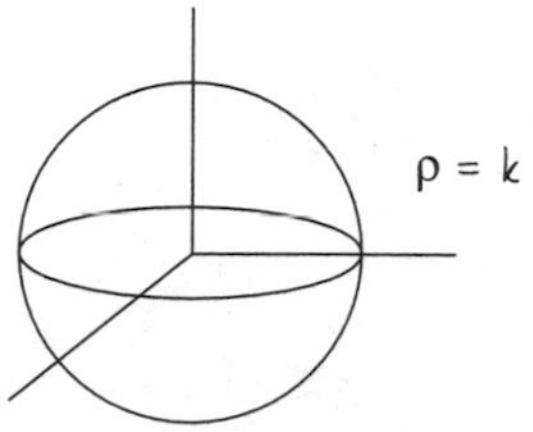

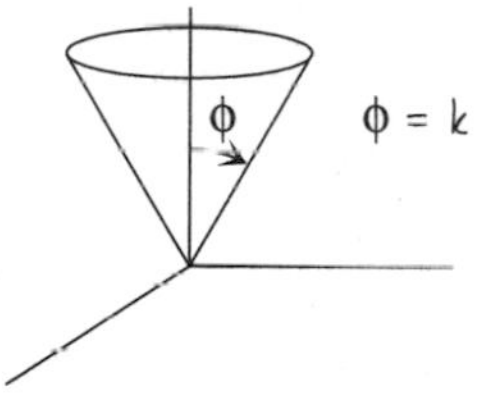

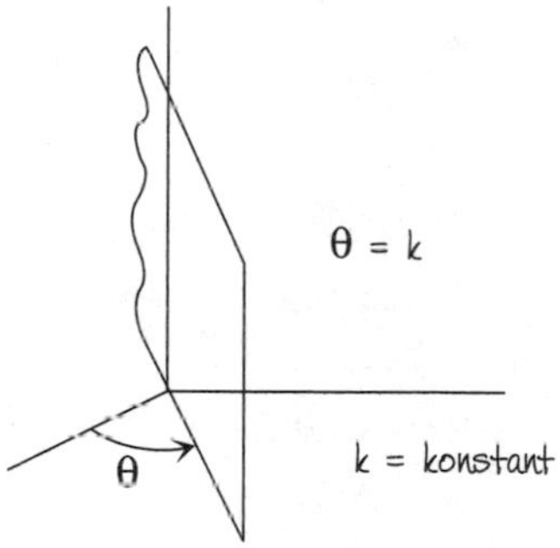

The next figure is the derivation of a little unit of volume. dzdydx becomes $\rho^2 \sin \phi \, d\rho \, d\phi \, d\theta$. Let's do a problem.

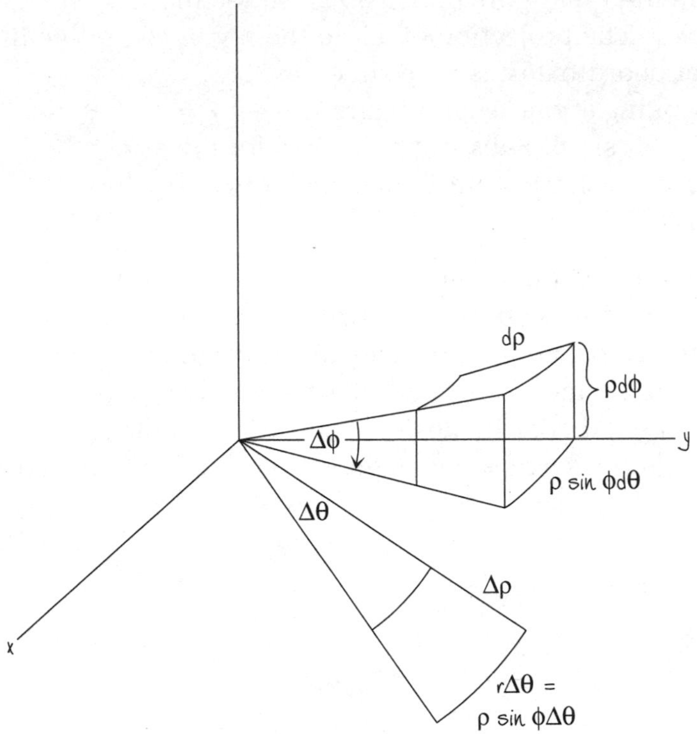

EXAMPLE 4—

Find the volume of a sphere or radius a. (We know the answer . . . V = (4/3)πa^3.)

$$8\int_{x=0}^{a}\int_{y=0}^{\sqrt{a^2-x^2}}\int_{z=0}^{z=\sqrt{a^2-x^2-y^2}} 1\ dz\ dy\ dx$$

$$=\int_{\theta=0}^{2\pi}\int_{\phi=0}^{\pi}\int_{\rho=0}^{a} \rho^2 \sin\phi\ d\rho\ d\phi\ d\theta$$

$$=\int_{\theta=0}^{2\pi}\int_{\phi=0}^{\pi} \frac{\rho^3}{3}\sin\phi\Big[_{\rho=0}^{a}\ d\phi\ d\theta = \int_{\theta=0}^{2\pi}\int_{\phi=0}^{\pi}\frac{a^3}{3}\sin\phi\ d\phi\ d\theta$$

$$=\int_0^{2\pi}\frac{a^3}{3}[-\cos\phi]_0^{\pi}\ d\theta = \int_{\theta=0}^{2\pi}\frac{2a^3}{3}\ d\theta = \frac{4}{3}\pi a^3$$

For completeness, we will include the formula for surface area. Since it is derived quite well in most books, we will state the theorem and do one problem on it. If $z = f(x,y)$, the surface area is

$$\iint_R \sqrt{f_x^2 + f_y^2 + 1}\, dA$$

where R is the region in the x-y plane that has $z = f(x,y)$ as a roof.

EXAMPLE 5—

Find the surface area of the part of the sphere $x^2 + y^2 + z^2 = 36$ above the x-y plane and inside the cylinder $x^2 + y^2 = 1$.

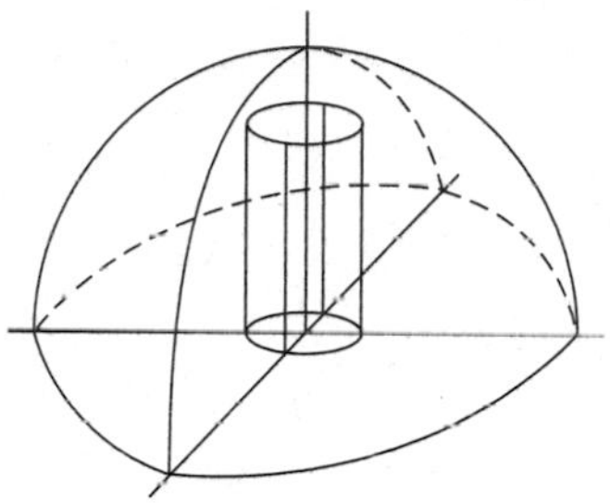

This problem has several tricks that are found in many of these problems.

$z = (36 - x^2 - y^2)^{1/2} = f(x,y)$

$f_x = -x/(36 - x^2 - y^2)^{1/2}$.

$$f_x^2 + f_y^2 + 1 = \frac{x^2}{36 - x^2 - y^2} + \frac{y^2}{36 - x^2 - y^2} + \frac{36 - x^2 - y^2}{36 - x^2 - y^2}$$

$$\sqrt{(f_x^2 + f_y^2 + 1)} = 6/\sqrt{(36 - x^2 - y^2)}$$

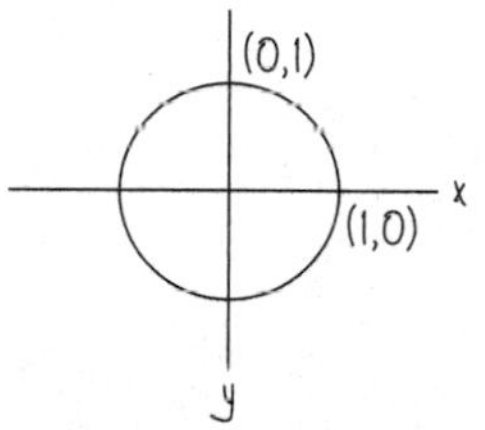

Change to polar!

$$\int_{y=-1}^{1} \int_{x=-\sqrt{1-y^2}}^{+\sqrt{1-y^2}} \frac{6}{\sqrt{36 - x^2 - y^2}}\, dx\, dy$$

$$\int_{\theta=0}^{2\pi} \int_{r=0}^{1} \frac{6r}{\sqrt{36 - r^2}}\, dr\, d\theta$$

$$= \int_{\theta=0}^{2\pi} -6\sqrt{36 - r^2}\Big[_{r=0}^{1}\, d\theta$$

$$= \int_{\theta=0}^{2\pi} 6[6 - \sqrt{35}]\, d\theta = 12\pi[6 - \sqrt{35}]$$

CHAPTER 7

INFINITE SEQUENCES

This topic brings some controversy. Some people think it is very difficult. Some think it is very easy. I believe if you understand the beginning, the rest of the chapter is not too bad.

DEFINITION

Sequence—A sequence of terms, technically, is a function for which the domain is the positive integers. Nontechnically, there is a first term called a_1 (read "ā sub-one," where the "one" is a subscript, not an exponent) denoting the first term, a_2 ("ā sub-two") denoting the second term, and so on. The notation for an infinite sequence is $\{a_n\}$.

Let us give some examples. We will list some sequences, write the first four terms, and then term number 100 by substituting 1, 2, 3, 4, . . . , 100 for n in a_n.

EXAMPLE 1—

$\{a_n\}$	1st	2nd	3rd	4th	100th
$\left\{\frac{n}{n+1}\right\}$	$\frac{1}{2}$	$\frac{2}{3}$	$\frac{3}{4}$	$\frac{4}{5}$	$\frac{100}{101}$
$\left\{\frac{(-1)^{n+1}(4n+1)}{n^2+1}\right\}$	$\frac{5}{2}$	$\frac{-9}{5}$	$\frac{13}{10}$	$\frac{-17}{17}$	$\frac{-401}{10001}$
$\{6\}$	6	6	6	6	6

DEFINITION (NONTECHNICAL)

We write $\lim_{n\to\infty} a_n = L$ if, the larger n gets, the closer a_n gets to L.

In this case, we say that $\{a_n\}$ converges to L (or has the limit L). If a_n goes to plus or minus infinity or does not go to a single number, then $\{a_n\}$ diverges (or has *no* limit).

EXAMPLE 2—

Find the limit of $\{(n + 9)/n^2\}$.

$a_n = (n + 9)/n^2 = 1/n + 9/n^2$. As n goes to infinity, both terms go to 0. Therefore, the sequence converges to 0.

EXAMPLE 3—

Find the limit of $\{(2n^2 + 3n + 2)/(5 - 7n^2)\}$.

Divide top and bottom of a_n by n^2. We get $[2 + (3/n) + (2/n^2)]/[(5/n^2 - 7)]$. As n goes to infinity, a_n goes to $2/(-7)$. The sequence has the limit $-2/7$.

NOTE

This should look very familiar. This is how we found horizontal asymptotes. Also note that we can use L'Hopital's rule.

EXAMPLE 4—

Find the limit of $\{a_n\} = \{\ln (n + 1) - \ln n\}$.

$$\lim_{n\to\infty} \ln(n+1) - \ln n = \lim_{n\to\infty} \ln[(n+1)/n]$$

$$= \lim_{n\to\infty} \ln(1 + 1/n) = \ln 1 = 0$$

The sequence converges to 0.

EXAMPLE 5—

Does $\{(-1)^n\}$ have a limit?

This sequence is $-1, +1, -1, +1, -1, \ldots$. There is no limit because the sequence does not go to one number.

DEFINITION

$\lim_{n\to\infty} a_n = L$ if, given an $\varepsilon > 0$, there exists an $N > 0$, such that if $n > N$, $|a_n - L| < \varepsilon$.

NOTE

It is not important that you know the technical definition of a limit to understand the rest of the chapter. But . . . at this point of your mathematical career, you should start understanding the background. It probably will help you later on. It would also be nice if you could see the beauty and the depth of this material—the beginnings of calculus. It truly is a wonderful discovery.

EXAMPLE 6—

Using ε, N, show $\lim_{n\to\infty} (2n + 5)/(n + 1) = 2$.

$$\left|\frac{2n+5}{n+1} - 2\right| = \left|\frac{2n+5}{n+1} - \frac{2(n+1)}{n+1}\right| = \left|\frac{3}{n+1}\right| < \varepsilon$$

provided $3/\varepsilon < n + 1$ or $3/\varepsilon - 1 < n$. We then choose N as the whole-number part of $3/\varepsilon - 1$.

The following theorems are used often. They are proved in many books and will only be stated here.

Let $\lim_{n\to\infty} a_n = L$, $\lim_{n\to\infty} b_n = M$; k = constant, f continuous. Then

1. $\lim_{n\to\infty} (a_n \pm b_n) = L \pm M$
2. $\lim_{n\to\infty} (a_n b_n) = LM$

3. $\lim_{n\to\infty} (a_n/b_n) = L/M \qquad M \neq 0$
4. $\lim_{n\to\infty} ka_n = kL$
5. $\lim_{n\to\infty} f(a_n) = f(L)$
6. $c_n \le d_n \le e_n$ and $\lim_{n\to\infty} c_n = \lim_{n\to\infty} e_n = P$. Then $\lim_{n\to\infty} d_n = P$.

EXAMPLE 7—

Show $\lim_{n\to\infty} (\sin n)/n = 0$.

Using part 6 above, $-1 \le \sin n \le 1$. So $-1/n \le (\sin n)/n \le 1/n$. As n goes to infinity, $-1/n$ and $1/n$ go to 0. Therefore, so does $(\sin n)/n$.

DEFINITION 1

An *increasing sequence* is one where $a_n < a_{n+1}$ for all n.

DEFINITION 2

A *nondecreasing sequence* is one where $a_n \le a_{n+1}$ for all n.

Similarly we can define *decreasing* and *nonincreasing.*

DEFINITION 3

A sequence is bounded if $|a_n| \le M$, some number M and all n.

Another theorem: Every bounded increasing (decreasing) sequence has a limit.

INFINITE SERIES

I know this is getting to be a drag, but it is essential to understand the terminology. This understanding will make the rest of the chapter *much* easier. I don't know why, but it really seems to.

DEFINITION

Partial sums—Given sequence $\{a_n\}$:

1st partial sum $S_1 = a_1$

2nd partial sum $S_2 = a_1 + a_2$

3rd partial sum $S_3 = a_1 + a_2 + a_3$

nth partial sum $S_n = a_1 + a_2 + a_3 + \cdots + a_n = \sum_{k=1}^{n} a_k$

The *infinite series* $a_1 + a_2 + a_3 + \cdots$ or $\sum_{k=1}^{\infty} a_k$ is said to *converge* to the sum S if $\lim_{n\to\infty} S_n = S$. If S does not exist, the series *diverges.*

EXAMPLE 8—

.767676. . . .

We can write this as an infinite series. .76 + .0076 + .000076 + ⋯. This is a geometric series (infinite). This is one of the few series we can find the exact sum of.

$S = a/(1 - r)$ $a = .76$ $r = .01$

$S = .76/(1 - .01) = 76/99$

More generally, the series $a + ar + ar^2 + ar^3 + \cdots$ converges to $a/(1 - r)$ if $|r| < 1$.

EXAMPLE 9—

$4 - 8 + 16 - 32 + \cdots$: $a = 4$, $r = -2$; diverges.

EXAMPLE 10—

$1 + 1 + 1 + 1 + 1 + \cdots$: $a = 1$, $r = 1$; diverges.

EXAMPLE 11—

$1 - 1 + 1 - 1 + 1 - \cdots$: $a = 1$, $r = -1$; diverges.

Note that test 1 implies divergence in these three examples.

EXAMPLE 12—

$$\sum_{k=1}^{\infty} \frac{1}{k(k+1)}$$

Using partial fractions

$$\frac{1}{k(k+1)} = \frac{1}{k} - \frac{1}{k+1}$$

Writing out the first few terms plus the n – 1 term plus the nth term, we get

$S_n = (1/1 - 1/2) + (1/2 - 1/3) + (1/3 - 1/4) + \cdots + [1/n - 1/(n+1)]$

Notice all the middle terms cancel out in pairs. So only the first and last terms remain:

$S_n = 1 - 1/(n+1) \qquad S = \lim_{n\to\infty} S_n = 1$

Again, this is one of the few sequences we can find the exact value for. (This is called a *telescoping series*—it collapses like one of those toy or portable telescopes.) From this point on, for almost all of the converging series, we will be able to tell that the series converges, but we won't be able to find its value. Later we will do some approximating.

EXAMPLE 13—

$$\sum_{k=1}^{\infty} \frac{4(2^k) + 5^k}{7^k}$$

After splitting, we get two geometric series:

$$S = \frac{4(2/7)}{1 - 2/7} + \frac{5/7}{1 - 5/7} = 8/5 + 5/2 = 41/10$$

THEOREM

If $\sum_{n=1}^{\infty} a_n = L$ and $\sum_{n=1}^{\infty} b_n = M$, then $\sum_{n=1}^{\infty} (ca_n + b_n) = cL + M$.

Now that we have an idea about what a sequence is and what an infinite series is (hopefully a *very good* idea), we would like to have some tests for when a series converges or diverges.

TEST I

It is necessary that $a_k \to 0$ for $\sum_{k=1}^{\infty} a_k$ to converge.

NOTE 1
If a_k does not go to 0, $\sum_{k=1}^{\infty} a_k$ diverges.

NOTE 2
If a_k does go to 0, and that is all we know, we know nothing.

EXAMPLE 14—

Tell whether $\sum_{k=1}^{\infty} k/(k+1)$ converges.

$k/(k+1)$ goes to 1. Therefore $\sum_{k=1}^{\infty} k/(k+1)$ diverges.

EXAMPLE 15—

The harmonic series $\sum_{k=1}^{\infty} 1/k$

Since $1/k$ goes to 0, we don't know if this series converges or diverges. We shall shortly show that the harmonic series diverges.

EXAMPLE 16—

The p_2 series $\sum_{k=1}^{\infty} 1/k^2$

Since $1/k^2$ goes to 0, again we can't tell. Shortly we shall show that the p_2 series converges.

TEST 2
Given a_k and $a_k > 0$, a_k goes to 0 for k big enough. Suppose we have a continuous function $f(x)$ such that $f(k) = a_k$. Then $\sum_{k=1}^{\infty} a_k$ and $\int_1^{\infty} f(x)\,dx$ either both converge or both diverge.

This theorem is easily explained by examples.

EXAMPLE 17—

Tell whether $\sum_{k=1}^{\infty} ke^{-k^2}$ converges or diverges.

The improper integral associated with $\sum_{k=1}^{\infty} ke^{-k^2}$ is $\int_1^{\infty} xe^{-x^2}\,dx$. Letting $u = -x^2$, $du = -2x\,dx$.

$$\int_1^{\infty} xe^{-x^2}\,dx = \lim_{b\to\infty} \int_1^{b} \frac{-2xe^{-x^2}}{-2}\,dx = \lim_{b\to\infty} (-½) \int_{-1}^{-b^2} e^u\,du$$

$$= \lim_{b\to\infty} -½(e^{-b^2} - e^{-1}) = \frac{1}{2e}$$

Since the improper integral converges, so does the infinite series.

NOTE 1

The value of the improper integral is not the value of the infinite series. But we can say the following: If the integral and the series together converge, then $\int_1^{\infty} f(x)\,dx \le \sum_{k=1}^{\infty} a_k \le a_1 + \int_1^{\infty} f(x)\,dx$.

The bounds on $\sum_{k=1}^{\infty} ke^{-k^2}$ are

$$\frac{1}{2e} \le \sum_{k=1}^{\infty} ke^{-k^2} \le \frac{1}{e} + \frac{1}{2e}.$$

NOTE 2

In this case, this is not too good an approximation.

We will get a better one if we take the fourth partial sum:

$$= \frac{1}{e} + \frac{2}{e^4} + \frac{3}{e^9} + \frac{4}{e^{16}}$$

The "error," the estimate on the rest of the terms, is that

$$\sum_{k=5}^{\infty} ke^{-k^2} \le a_5 + \int_5^{\infty} xe^{-x^2}\,dx = \frac{5}{e^{25}} + \frac{1}{2e^{25}}\ 1.4 \times 10^{-11}$$

This is more accuracy than you will probably ever need!!! Lots of things you cannot even integrate.

EXAMPLE 18—

The harmonic series $\sum_{k=1}^{\infty} 1/k$ diverges.

$\lim_{b\to\infty} \int_1^{b} 1/x\,dx = \ln b$, which goes to infinity as b goes to infinity.

EXAMPLE 19—

The p_2 series $\sum_{k=1}^{\infty} 1/k^2$ converges.

$\int_1^b 1/x^2\,dx = -1/b + 1$. Since $-1/b$ goes to 0 as b goes to infinity, this improper integral converges. So does the p_2.

EXAMPLE 20—

$$\sum_{k=1}^{\infty} 1/k^p$$

If $p > 1$, it converges, and if $p \leq 1$, it diverges. Just use the integral test. It's easy.

TEST 3

The comparison test. Given $\sum_{k=1}^{\infty} a_k$, $\sum_{k=1}^{\infty} b_k$ where $0 < a_k \leq b_k$,

1. If $\sum_{k=1}^{\infty} b_k$ converges, so does $\sum_{k=1}^{\infty} a_k$.
2. If $\sum_{k=1}^{\infty} a_k$ diverges, so does $\sum_{k=1}^{\infty} b_k$.

Let us talk through part 1. The second part can be shown to be equivalent. The partial sums S_n of the $\sum_{k=1}^{\infty} b_k$ series are uniformly bounded because the first N terms are bounded by their maximum and the rest are bounded by $L + \varepsilon$. Therefore the partial sums of the $\sum_{k=1}^{\infty} a_k$ series also are bounded, being respectfully smaller than those of $\sum_{k=1}^{\infty} b_k$. Moreover, since $a_k > 0$, then the partial sums of the a_k form an increasing sequence. Now we have an increasing bounded sequence that has a limit. Therefore, $\sum_{k=1}^{\infty} a_k$ converges.

EXAMPLE 21—

Examine $\sum_{k=1}^{\infty} 1/(4 + k^4)$.

$1/(4 + k^4) < 1/k^4$. $\sum_{k=1}^{\infty} 1/k^4$ converges by the previous example. Since the given series is smaller termwise than a convergent series, it must converge by the comparison test.

EXAMPLE 22—

Examine $\sum_{k=1}^{\infty} (2 + \ln k)/k$.

$\sum_{k=1}^{\infty} (2 + \ln k)/k > \sum_{k=1}^{\infty} 2/k$ = twice a divergent series (the harmonic). Since the given series is larger termwise than a divergent series, the given series must diverge.

TEST 4

This is the limit comparison test. Given $\sum_{k=1}^{\infty} a_k$ and $\sum_{k=1}^{\infty} b_k$, $a_k \geq 0$, $b_k \geq 0$. If $\lim_{k\to\infty} (a_k/b_k) = r$, where r is any positive number, both series converge, or both diverge.

EXAMPLE 23—

$$\sum_{k=1}^{\infty} 3/(5k^4 + 4)$$

Let us compare this series with $\sum_{k=1}^{\infty} 1/k^4$.

Divide top and bottom by k^4.

$$\lim_{k\to\infty} [3/(5k^4 + 4) \text{ divided by } 1/k^4]$$

$$= \lim_{k\to\infty} \frac{3k^4}{5k^4 + 4}$$

$$= \lim_{k\to\infty} \frac{3}{5 + (4/k^4)} = 3/5$$

Since the limit is a positive number, both series do the same thing. Since $\sum_{k=1}^{\infty} 1/k^4$ converges, so does $\sum_{k=1}^{\infty} 3/(5k^4 + 4)$.

TEST 5 (RATIO TEST)

Given $a_k \geq 0$, $\lim_{k\to\infty} (a_{k+1}/a_k) = r$. If $r > 1$, it diverges. If $r < 1$, it converges. If $r = 1$, use another test.

EXAMPLE 24—

Examine $\sum_{k=1}^{\infty} k^2/5^k$.

$a_{k+1}/a_k = (k+1)^2/5^{k+1}$ divided by $k^2/5^k$

$$= \frac{(k+1)^2}{5^{k+1}} \times \frac{5^k}{k^2} = \frac{k^2 + 2k + 1}{5k^2}$$

$$\lim_{k\to\infty} \frac{a_{k+1}}{a_k} = \lim_{k\to\infty} \frac{1 + \frac{2}{k} + \frac{1}{k^2}}{5} = \frac{1}{5} < 1 \qquad \sum_{k=1}^{\infty} k^2/5^k$$

The series converges.

EXAMPLE 25

Examine $\sum_{k=1}^{\infty} 7^k/k!$

NOTE

6! means 6(5)(4)(3)(2)(1).

ALSO NOTE

$(k + 1)! = (k + 1)(k!)$. That is, 10! = 10(9!), and so on.

Let us again use the ratio test.

$a_{k+1}/a_k = 7^{k+1}/(k+1)!$ divided by $7^k/k!$

$$= \frac{7^{k+1}}{(k+1)!} \times \frac{k!}{7^k} = \frac{7}{k+1}$$

$$\lim_{k\to\infty} \frac{7}{k+1} = 0 < 1 \qquad \sum_{k=1}^{\infty} 7^k/k!$$

The series converges.

EXAMPLE 26

Let's now look at $\sum_{k=1}^{\infty} k^k/k!$

This is a little trickier than most. Again, we use the ratio test.

$a_{k+1}/a_k = (k+1)^{k+1}/(k+1)!$ divided by $k^k/k!$

$$= \frac{(k+1)^{k+1}}{(k+1)!} \times \frac{k!}{k^k} = \frac{(k+1)(k+1)^k k!}{(k+1)k!k^k} = \frac{(k+1)^k}{k^k}$$

$$= (1 + 1/k)^k$$

Since the $\lim_{k\to\infty} [1 + (1/k)]^k = e > 1$, $\Sigma_{k=1}^{\infty} k^k/k!$ diverges.

EXAMPLE 27—

In order to show the third part of the previous theorem, you should apply the ratio test to both the harmonic series and the p_2 series. Both give a ratio of 1. The first series diverges, and the second converges. So, if the ratio is 1, we must indeed use another test.

TEST 6

(Root test). Given $\Sigma_{k=1}^{\infty} a_k$. Take $\lim_{k\to\infty} (a_k)^{1/k} = r$ $(a_k \geq 0)$. If $r > 1$, it diverges. If $r < 1$, it converges. If $r = 1$, use another test.

NOTE

To show the third part ($r = 1$), we would again use the harmonic and p_2 series. Let us give examples for the first two parts ($r > 1$ and $r < 1$).

EXAMPLE 28—

$$\sum_{k=1}^{\infty} 3^k/k^k$$

Take $(3^k/k^k)^{1/k} = 3/k$. $\lim_{k\to\infty} (3/k) = 0 < 1$. So the series converges.

EXAMPLE 29—

$$\sum_{k=1}^{\infty} 2^k/k^2$$

Take $(2^k/k^2)^{1/k} = 2/k^{2/k}$. $\lim_{k\to\infty} 2/k^{2/k} = 2/1 = 2$ $(\lim_{k\to\infty} k^{2/k} = \lim_{k\to\infty} e^{2(\ln k)/k} = e^{2(0)} = 1)$. Since $2 > 1$, the series diverges.

Up to this time, we have dealt exclusively with positive terms. Now we will deal with infinite series that have terms that alternate from positive to negative. We will assume the first term is positive. The notation will be as follows: alternating series $\sum_{k=1}^{\infty} (-1)^{k+1} a_k$, where all a_k are positive.

TEST 7

Given an alternating series where (a) $0 < a_{k+1} \le a_k$, $k = 1, 2, 3, 4, \ldots.$ and (b) $\lim_{k\to\infty} a_k = 0$, the series converges to S and $S \le a_1$.

In clearer English, the *only* thing you must do to show an alternating series converges is to show the terms go to zero. (If only all series were that easy!)

EXAMPLE 30—

Alternating harmonic $\sum_{k=1}^{\infty} ((-1)^{k+1})/k$ converges since the terms go to 0.

DEFINITION

Absolutely convergent—A series $\sum_{k=1}^{\infty} a_k$ converges absolutely if $\sum_{k=1}^{\infty} |a_k|$ converges.

NOTE

If a series converges absolutely, it converges.

DEFINITION

Conditionally convergent—A series $\sum_{k=1}^{\infty} a_k$ converges conditionally if it converges but $\sum_{k=1}^{\infty} |a_k|$ diverges.

NOTE 1

If we have an alternating series and want to show that it converges conditionally, we only have to show its terms go to zero. To find out whether it is absolutely convergent, we must use some other test.

NOTE 2

There are three possibilities for an alternating series: it diverges, converges conditionally, or converges absolutely.

Let us look at three alternating series.

EXAMPLE 31—

Let us look at $\sum_{k=1}^{\infty} (-1)^{k+1}/(2k+1)$. This series converges conditionally since (1) the terms go to zero, but (2) using the limit comparison test with the harmonic series, the positive series behaves as the harmonic series and diverges.

EXAMPLE 32—

What about the series $\sum_{k=1}^{\infty} (-1)^{k+1}/(k^2+1)$? This series converges absolutely by comparing to the p_2 series.

EXAMPLE 33—

$\sum_{k=1}^{\infty} (-1)^{k+1}k^2/(k^2+6)$ diverges since the terms don't go to 0.

DEFINITION

Region of convergence—We have an infinite series whose terms are functions of x. The set of all points x for which the series converges is called the *region of convergence*.

Now let's get back to the series with x's in them. Series of this type are usually done with the *ratio test*. This is to find the region of convergence. Then, you will test both the left and right end points. There are three tests in all.

EXAMPLE 34—

$$\sum_{k=1}^{\infty} x^k/k$$

Using the ratio test,

$|a_{k+1}/a_k| = |x^{k+1}/(k+1)|$ divided by $|x^k/k|$

$$= \left|\frac{x^{k+1}}{k+1}\right| \times \left|\frac{k}{x^k}\right| = \left|\frac{kx}{k+1}\right| \cdot \lim_{k\to\infty} \left|\frac{kx}{k+1}\right| = |x|$$

So the region of convergence is $|x| < 1$ or $-1 < x < 1$.

Let us test both 1 and −1 by substituting those values into the original series. $x = 1$ gives us $\sum_{k=1}^{\infty} 1^k/k$ or $\sum_{k=1}^{\infty} 1/k$, the harmonic series that diverges. $x = -1$ gives us $\sum_{k=1}^{\infty} (-1)^k/k$, the alternating harmonic series, or rather the negative of the alternating harmonic series, since the first term is negative. We know this converges. Therefore, the region of convergence is $-1 \le x < 1$.

IMPORTANT NOTE

When you test the end points, anything is possible. Both ends could converge, both could diverge, the left could converge and not the right, or the right could converge but not the left.

EXAMPLE 35—

Let's look at $\sum_{k=1}^{\infty} (x-4)^k/3^k$.

This is a geometric series that converges for

$$|r| = \left|\frac{x-4}{3}\right| < 1.$$

Thus, the region of convergence is $|x-4| < 3$ or $1 < x < 7$. Test $x = 1$ and substitute into the original series. We get

$$\sum_{k=1}^{\infty} \frac{(-3)^k}{3^k} \text{ or } \sum_{k=1}^{\infty} (-1)^k = -1 + 1 - 1 + 1 \cdots$$

which diverges (Example 11). For $x = 7$, we get

$$\sum_{k=1}^{\infty} \frac{3^k}{3^k} = 1 + 1 + 1 + 1 + \cdots$$

which diverges (Example 10).

EXAMPLE 36—

$\sum_{k=1}^{\infty} x^k/k!$, a nice one.

$|a_{k+1}/a_k| = |x^{k+1}/(k+1)|$ divided by $|x^k/k!|$

$$= \left|\frac{x^{k+1}}{(k+1)!}\right| \times \left|\frac{k!}{x^k}\right| = \left|\frac{xk!}{(k+1)k!}\right| = \left|\frac{x}{k+1}\right|$$

$$\lim_{k\to\infty}\left|\frac{x}{k+1}\right| = 0.$$

This says no matter what x is, the limit will always be less than 1. The region of convergence is *all real numbers*.

EXAMPLE 37—

$$\sum_{k=1}^{\infty} (k+3)!x^k$$

$|a_{k+1}/a_k| = |(k+4)!x^{k+1}|$ divided by $|(k+3)!x^k|$

$$= \left|\frac{(k+4)(k+3)!x^{k+1}}{(k+3)!x^k}\right| = |(k+4)x|$$

$\lim_{k\to\infty} |(k+4)x| \to \infty$ except if $x = 0$. The region of convergence is just the point $x = 0$.

Which Test to Use

After finishing the original draft of this book but before running off copies, I finished student-testing this section on infinite series. It became absolutely clear that the following list is necessary.

1. Always see if the terms go to zero first. If they don't, the series diverges. If the terms go to zero, the series at least converges conditionally if it alternates.

2. Use the integral test if the infinite series looks like an integral you have done. By this time, you should have so many integrals you should be familiar with and/or sick of them.

3. Don't use the integral test if you can see an easier one or if there is a factorial symbol.
4. My favorite is the ratio test. Always try the ratio test if there is a factorial or an x in the problem. Also try the ratio test if there is something to a power of k, such as 2^k or k^k.
5. Use the limit comparison test or the comparison test if the series looks like one you know like the harmonic series, p_2 series, and so on. Use the comparison if the algebra is not too bad. Use the limit comparison if the algebra looks really terrible or even semiterrible.
6. Use the root test if there is at least one term with k in the exponent and no factorial in the problem.
7. If there is a series where there are a lot of messy-looking terms multiplying each other, the ratio test is probably the correct one.
8. Sometimes you may not be able to tell the terms go to zero. The ratio test may give absolute convergence or divergence immediately.
9. Practice factorial. It is new to most of you. Once again, note that $(2n + 1)! = (2n + 1)(2n)! = (2n + 1)(2n)(2n - 1)!$, that is, $7! = 7(6!) = 7(6)(5!)$. Study factorial!!!!!
10. Most of all, do a lot of series testing. You will get better if you practice. The nice part is that the problems are mostly very short.

A PREVIEW OF POWER SERIES

We would like to have a polynomial approximation of a function in the vicinity of a given point. Polynomials are very easy to work with. They can be integrated easily, while many functions can't be integrated at all. Exact answers are usually not needed, since we do not live in a perfect world.

We therefore have Taylor's theorem, which gives us a polynomial that approximates f(x) for every x approximately equal to a; the closer x is to a, the better the approximation for a given length. The "meat" of the theorem is a formula for the remainder, or error, when you replace the function by the polynomial. This is necessary so that you know how close your answer is.

Taylor's Theorem

1. $f^{(n+1)}(x)$, [n + 1 derivatives], continuous on some interval, I, where x is in the interval.

2. a is any number in the interval I, usually its midpoint.

3. $$S_n(x) = f(a) + \frac{f'(a)(x-a)}{1!} + \frac{f''(a)(x-a)^2}{2!} + \frac{f'''(a)(x-a)^3}{3!} + \cdots + \frac{f^{(n)}(a)(x-a)^n}{n!}$$

   NOTE

   $S_n(x)$ is the sum of the polynomial terms up to term of degree n.

4. The remainder $R_n(x) = f(x) - S_n(x)$ for all x in I.

Then there is a point w in I—w is between a and x—such that

$$R_n(x) = \frac{f^{(n+1)}(w)(x-a)^{n+1}}{(n+1)!}$$

Let us give three examples worked out all the way.

EXAMPLE 38—

$f(x) = e^x$; $a = 0$.

Write a polynomial of degree 2. Write the remainder. Find the approximate value for $e^{.2}$ and estimate the maximum error from the actual value of $e^{.2}$.

This sounds like a lot of work, but, as we will see, this process for e^x (e doesn't stand for "easy," but it should) is really quite short.

$f(x) = e^x \qquad f'(x) = e^x \qquad f''(x) = e^x \qquad f'''(x) = e^x$

$f(0) = f'(0) = f''(0) = 1 \qquad f'''(w) = e^w$

$$f(x) = \underbrace{f(a) + \frac{f'(a)(x-a)}{1!} + \frac{f''(a)(x-a)^2}{2!}}_{S_2(x)} + \underbrace{\frac{f'''(w)(x-a)^3}{3!}}_{R_2(x)}$$

$e^x = 1 + x/1 + x^2/2 + e^w x^3/6$

Therefore, $S_2(.2) = 1 + .2 + (.2)^2/2 = 1.22$ and $R_2(.2) = e^w(.2)^3/6$, where w is between 0 and .2. Because e^x is an increasing function, $e^w < e^{.2} < e^{.5} < (3)^{1/2} < 2$ (being very wasteful). Therefore, $R_2(.2) < 2(.2)^3/6 \approx .00267$, the maximum error.

This is a pretty good approximation. Remember, we were really rough-estimating the error, and this is only a polynomial of degree 2.

EXAMPLE 39—

Let us do the same for ln (1 + x), polynomial degree 3, $a = 0$, $x = 1$, estimate the error for ln 1.1.

$f(x) = \ln(x + 1) \qquad f'(x) = (x + 1)^{-1} \qquad f''(x) = (-1)(x + 1)^{-2}$

$f'''(x) = (-1)(-2)(x + 1)^{-3} \qquad f''''(x) = (-1)(-2)(-3)(x + 1)^{-4}$

$f(0) = 0 \qquad f'(0) = 1 \qquad f''(0) = -1 \qquad f'''(0) = 2$

$f''''(w) = -6(w + 1)^{-4}$

$$f(x) = f(a) + f'(a)(x-a) + \frac{f''(a)(x-a)^2}{2!} + \frac{f'''(a)(x-a)^3}{3!} + \frac{f''''(w)(x-a)^4}{4!}$$

$$\ln(1+x) = 0 + x - 1x^2/2 + 2x^3/6 - 6(w+1)^{-4}x^4/4!$$

Therefore $S_3(x) = x - x^2/2 + x^3/3$ and $\ln(1.1) = .1 - (.1)^2/2 + (.1)^3/3 = .098333\cdots$. $R_3(x) = -x^4/4(w+1)^4$.

Let's estimate. $0 < w < .1$, so $1 < w + 1 < 1.1$. Consequently, $1/1.1 < 1/(w+1) < 1$; $|R_3(.1)| = (.1)^4/4(w+1)^4 < .1^4/4 = .000025$. Not bad!!

EXAMPLE 40—

Let's do the same for $f(x) = \sin x$ except let $a = 30° = \pi/6$ with a polynomial of degree 3, and $x = 32°$. Hold on to your hats, 'cause this is pretty messy.

$f(x) = \sin x \qquad f'(x) = \cos x \qquad f''(x) = -\sin x$

$f'''(x) = -\cos x \qquad f^{(4)}(x) = \sin x \qquad f(\pi/6) = ½$

$f'(\pi/6) = 3^{1/2}/2 \qquad f''(\pi/6) = ½ \qquad f'''(\pi/6) = 3^{1/2}/2$

$f^{(4)}(w) = \sin w$

$$f(x) = f(a) + f'(a)(x-a) + \frac{f''(a)(x-a)^2}{2!} + \frac{f'''(a)(x-a)^3}{3!} + \frac{f^{(4)}(w)(x-a)^4}{4!}$$

$$\sin x = ½ + 3^{1/2}/2\,\frac{(x-\pi/6)}{1!} - \frac{½(x-\pi/6)^2}{2!} - \frac{(3^{1/2}/2)(x-\pi/6)^3}{3!} + \frac{(\sin w)(x-\pi/6)^4}{4!}$$

Now $x = 32° = 32\pi/180$. So

$$x - \pi/6 = \frac{32\pi}{180} - \frac{30\pi}{180} = \frac{\pi}{90} = .035$$

$$\text{In } 32° = ½ + (3^{1/2}/2)(.035) - \frac{½(.035)^2}{2} - \frac{(3^{1/2}/2)(.035)^3}{6}$$
$$= .5299985,$$

the accuracy is not guaranteed, since $\pi/90$ should be more places.

Actually, I couldn't bear this inaccuracy. So $\pi/90$ is .0349066 on my calculator, and if I hit the right buttons, sin 32° = .5299195 (and my calculator is OK). The remainder is $(\sin w)(.0349066)^4/4!$ We know sin w is less than 1, so the remainder is less than $(.0349066)^4/4! = 6.1861288 \times 10^{-8}$.

Even with these limited examples, we see different series get more accurate results with the same number of terms. This can be studied in great detail.

Also, it is very convenient to know certain power series by heart. We will list the most important together with region of convergence.

e^x	all reals	$1 + x + x^2/2! + x^3/3! + \cdots + x^n/n! + \cdots$
	$n = 0, 1, 2, 3, \ldots$	
sin x	all reals	$x/1! - x^3/3! + x^5/5! - x^7/7! \cdots (-1)^k x^{2k+1}/(2k+1)! \cdots$
	$k = 0, 1, 2, 3, \ldots$	
cos x	all reals	$1 - x^2/2! + x^4/4! - x^6/6! \cdots (-1)^k x^{2k}/(2k)! \cdots$
	$k = 0, 1, 2, 3, \ldots$	
ln (x + 1)	$-1 < x \le 1$	$x - x^2/2 + x^3/3 - x^4/4 \cdots (-1)^{k+1} x^k/k \cdots$
	$k = 1, 2, 3, 4, \ldots$	

Here are some more.

$1/(1 - x)$	$-1 < x < 1$	$1 + x + x^2 + x^3 + \cdots + x^k + \cdots$
$k = 0, 1, 2, 3, \ldots$		
$1/(x + 1)$	$-1 < x < 1$	$1 - x + x^2 - x^3 + \cdots (-1)^k x^k$
$k = 0, 1, 2, 3, \ldots$		

Binomial $f(x) = (1 + x)^p \quad -1 \le x < 1$

$1 + px + p(p - 1)x^2/2! + p(p - 1)(p - 2)x^3/3!$

$+ \cdots + p(p - 1)\cdots[p - (n - 1)]x^n/n! \cdots \quad n = 0, 1, 2, 3, \ldots$

Finally there are theorems found in many books that give the conditions under which you can add, subtract, multiply, divide, differentiate, and integrate infinite series. We can amaze ourselves by the number of functions we can approximate.

EXAMPLE 41—

Find the infinite series for cosh x.

$\cosh x = (e^x + e^{-x})/2$

$e^x = 1 + x + x^2/2! + x^3/3! + x^4/4! + \cdots$

Substitute –x for x. $e^{-x} = 1 - x + x^2/2! - x^3/3! + x^4/4! - \cdots$

Add and divide by 2. $\cosh x = 1 + x^2/2! + x^4/4! + x^6/6! + \cdots$

Pretty neat, eh?! More to come.

EXAMPLE 42—

$\int_0^1 e^{-x^2}\,dx$, four terms:

$e^x = 1 + x + x^2/2! + x^3/3!$

$e^{-x^2} = 1 - x^2 + x^4/2 - x^6/6$

$\int_0^1 e^{-x^2}\,dx = x - x^3/3 + x^5/10 - x^7/42$

$= 1 - 1/3 + 1/10 - 1/42 = 78/105$

Not too shabby. More to come.

EXAMPLE 43—

The series for $x/(1 + x)^2$:

Differentiating, we get: $1/(1 + x) = 1 - x + x^2 - x^3 + x^4 + \cdots$

$-1/(1 + x)^2 = -1 + 2x - 3x^2 + 4x^3 - 5x^4 + \cdots$

Multiply by −x; our result is:

$x/(1 + x)^2 = x - 2x^2 + 3x^3 - 4x^4 + 5x^5 - \cdots$

When mathematicians do things like this, you tend to believe that mathematics can do everything and anything. However, this is not true. However, the best is yet to come!!!!!!

We can derive every property of the sine and cosine using infinite series, never, *never* mentioning triangles or angles. Amazing, huh?! Given $\sin x = x - x^3/3! + x^5/5! \ldots$ and $\cos x = 1 - x^2/2! + x^4/4! - x^6/6!. \ldots$ How about cos 2x? $\cos 2x = 1 - 2x^2 + (2/3)x^4 - (4/45)x^6 \ldots$ [2x for x in cos x].

How about $\sin^2 x + \cos^2 x = 1$?

$$\begin{aligned}\sin^2 x + \cos^2 x &= (x - x^3/6 + x^5/120 \cdots)(x - x^3/6 + x^5/120\cdots)\\ &\quad + (1 - x^2/2 + x^4/24 - x^6/720\cdots)\\ &\quad \times (1 - x^2/2 + x^4/24 - x^6/720\cdots)\\ &= x^2 - x^4/3 + 2x^6/45\cdots + 1 - x^2 + x^4/3 - 2x^6/45\cdots\\ &= 1 + 0 + 0 + 0\cdots = 1\end{aligned}$$

How about tan x? Well we would like $\tan x = \sin x/\cos x$.

$$\begin{array}{r} x + x^3/3 + 2x^5/15 + 17x^7/315\cdots = \tan x \\ 1 - x^2/2 + x^4/24 - x^6/72\cdots \overline{\big)\, x - x^3/6 + x^5/120 - x^7/540\cdots} \end{array}$$

How about derivatives? If f(x) = sin x, we want f′(x) = cos x.

$$\sin x = x - x^3/6 + x^5/120 - x^7/5040\cdots$$

$$\begin{aligned}(\sin x)' &= 1 - 3x^2/6 + 5x^4/120 - 7x^6/5040\cdots\\ &= 1 - x^2/2 + x^4/24 - x^6/720\cdots = \cos x\end{aligned}$$

We could, of course, integrate cos x term by term and get sin x. We could go on and on, getting every property of the sine and cosine and all other trig functions totally without angles or triangles. The beauty of these power series is that they are limits of polynomials and are easy to deal with. Yet there are things that are even more powerful in mathematics.

CHAPTER 8

VECTOR FIELDS

The last set of topics are found in various courses: Calc III, Advanced Calculus, Vector Calculus, and so on. When they are done in Calc III, they usually are done without much depth. These topics will be presented without proofs the way that this whole series was intended—so that you understand enough to utilize the book and your instructor better. OK, let's go.

DEFINITION

Vector field—Let D be a subset of 2D. F is a vector function if it assigns to every point in D a vector. F = M(x,y)i + N(x,y)j.

EXAMPLE 1—

Draw the vector field **F** = ¼yi – ½xj. Let's do this somewhat systematically.

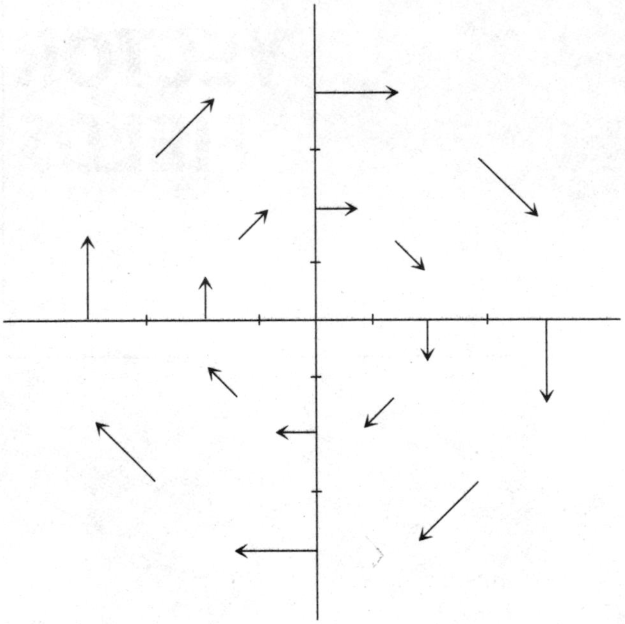

Point	Vector
(0,0)	⟨0,0⟩
(2,0)	⟨0,–1⟩
(4,0)	⟨0,–2⟩
(–2,0)	⟨0,1⟩
(–4,0)	⟨0,2⟩
(0,2)	⟨½,0⟩
(0,4)	⟨1,0⟩
(2,2)	⟨½,–1⟩
(4,4)	⟨1,–2⟩
(–2,–2)	⟨–½,1⟩
(–4,–4)	⟨–1,2⟩
(2,–2)	⟨–½,–1⟩
(4,–4)	⟨–1,–2⟩
(–2,2)	⟨½,1⟩
(–4,4)	⟨1,2⟩
(0,–2)	⟨–½,0⟩
(0,–4)	⟨–1,0⟩

NOTE

The same can be done in 3D.

F = M(x,y,z)i + N(x,y,z)j + P(x,y,z)k

Depending on the book, the following topic is done two ways: vectorwise and parametrically. Let's do it both ways.

DEFINITION

Smooth (can you believe we have to define smooth??!!!)—

Let r be defined on [a,b]: r = ⟨x(t),y(t)⟩. r is smooth *if*

1. r is continuous on [a,b]
2. so is r′(t)
3. r′(t) is never equal to 0. [x′(t),y′(t)] cannot both be zero at the same value of t.

A curve is smoooooth (curve C) if it has one, at least, position vector r.

A curve is piecewise smooth if it is the sum of a finite number of smooth curves.

$C = C_1 + C_2 + C_3 + C_4$

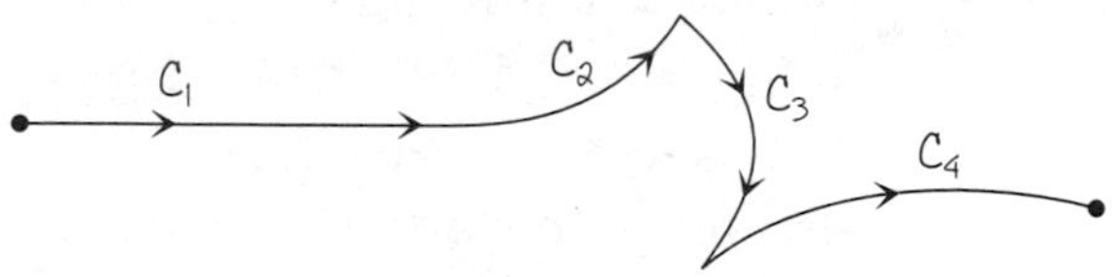

Let's do a bunch of examples. To check them out, all you have to do is plug in some t values.

Picture	Parameters	Vectors
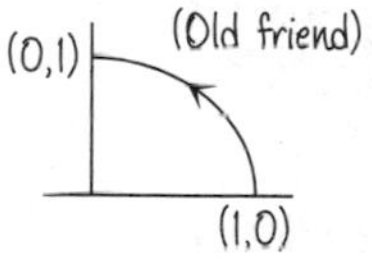	$x = \cos t \quad y = \sin t$	$r = \langle \cos t, \sin t \rangle$
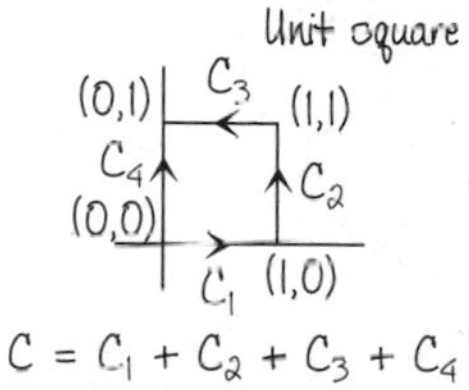	$C_1 \quad x = t, y = 0 \quad 0 \le t \le 1$ $C_2 \quad x = 1, y = t \quad 0 \le t \le 1$ $C_3 \quad x = 1 - t, y = 1 \quad 0 \le t \le 1$ $C_4 \quad x = 0, y = 1 - t \quad 0 \le t \le 1$	$r_1 = \langle t,0 \rangle \quad 0 \le t \le 1$ $r_2 = \langle 1,t \rangle$ $r_3 = \langle 1 - t,1 \rangle$ $r_4 = \langle 0,1 - t \rangle$
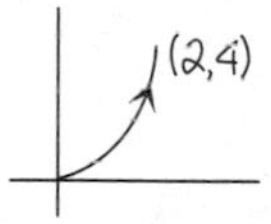	$x = t \quad y = t^2 \quad 0 \le t \le 2$	$r = \langle t,t^2 \rangle$
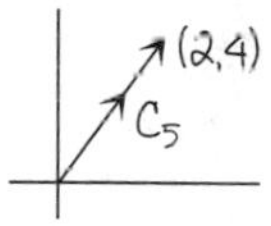	$x = t \quad y = 2t \quad 0 \le t \le 2$	$\langle t,2t \rangle \quad 0 \le t \le 2$
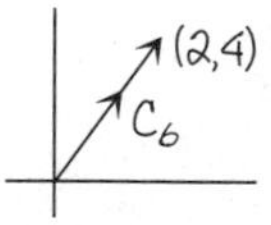	$x = 2t \quad y = 4t \quad 0 \le t \le 1$	$\langle 2t,4t \rangle \quad 0 \le t \le 1$
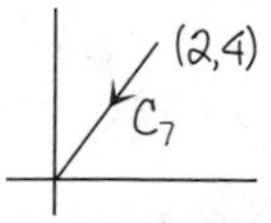	$x = 2 - t \quad y = 4 - 2t \quad 0 \le t \le 2$	$\langle 2 - t,4 - 2t \rangle \quad 0 \le t \le 2$

NOTE

C_5, C_6 have different representations, but they give exactly the same points in exactly the same order. We write $C_5 = C_6$.

NOTE ALSO

C_6 and C_7 give exactly the same points but in exactly the opposite order. We will write $C_6 = -C_7$.

A curve is *simple* if it does not cross itself. A curve is *closed* if the initial and end points are the same. The unit square we did on the last page is a simple, closed curve.

Where is this leading?? We will define the line integral in the next chapter.

CHAPTER 9

LINE INTEGRALS

We will define the line integral of F over C. But first, let's do lots of intro.

Given curve C with **r** defined on [a,b] and vector field F, C is in some set R where F is defined.

$\mathbf{F} = M(x,y)i + N(x,y)j$; $\mathbf{r} = x(t)i + y(t)j$;

$\mathbf{dr} = dxi + dyj = x'(t)dt\ i + y'(t)dt\ j$

Then the line integral $\int_C F \cdot dr$ is defined the following way:

$$\int_C \mathbf{F} \cdot \mathbf{dr} = \int_a^b \mathbf{F}(r(t)) \cdot \mathbf{r'(t)}\ dt$$

$$\left(= \int_a^b M(x(t),y(t))x'(t)\ dt + N(x(t),y(t))y'(t)\ dt\right)$$

if the right side exists.

EXAMPLE 2—

$$\int_C \mathbf{F} \cdot \mathbf{dr}$$

Find the line integral if $C = C_1 + C_2$ and $\mathbf{F} = x^2 i + xyj$.

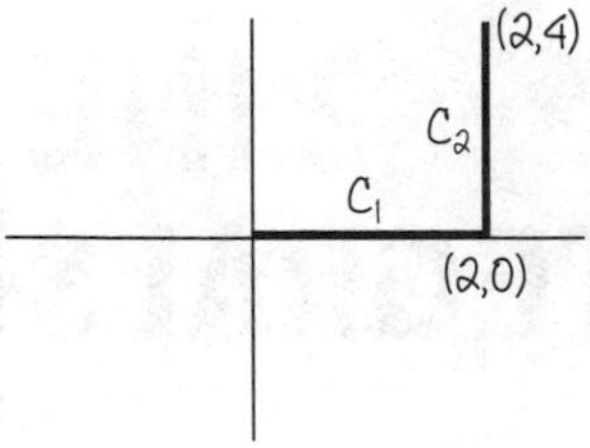

Let us parametrize C_1, C_2, F and dr.

C_1: $x = t$, $y = 0$; $dx = dt$, $dy = 0$; $F = x^2i + xyj = t^2i + 0j$; $0 \le t \le 2$.

$$\int_{C_1} \mathbf{F} \cdot \mathbf{dr} = \int_0^2 t^2\, dt = t^3/3 \Big[_0^2 = 8/3$$

C_2: $x = 2$, $y = t$; $dx = 0$, $dy = dt$; $\mathbf{F} = 4i + 2tj$; $0 \le t \le 4$.

$$\int \mathbf{F} \cdot \mathbf{dr} = \int (4i + 2tj) \cdot (0i + dtj) = \int_{t=0}^4 2t\, dt = t^2 \Big[_0^4 = 16$$

$$\int_C = \int_{C_1} + \int_{C_2} = 8/3 + 16 = 18\tfrac{2}{3}$$

Let's try another problem. Same initial point, same final point, same **F**, but a different path.

EXAMPLE 3A—

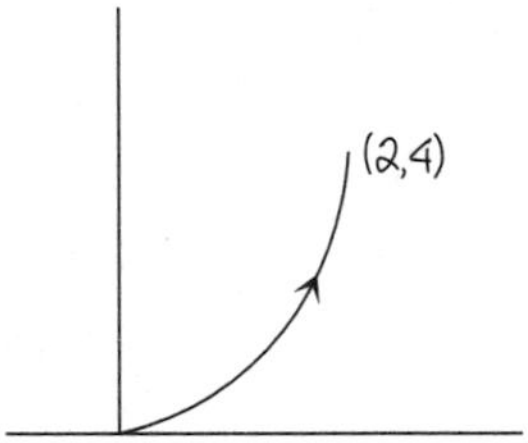

$\mathbf{F} = \langle x^2, xy \rangle$

C_3: $x = t$, $y = t^2$, $0 \le t \le 2$

$\mathbf{F} = \langle t^2, t^3 \rangle$ $\mathbf{dr} = \langle dt, 2tdt \rangle$

$$\int \langle t^2, t^3 \rangle \cdot \langle dt, 2tdt \rangle = \int_{t=0}^2 (t^2 + 2t^4)\, dt$$

$$= t^3/3 + 2t^5/5 \Big[_0^2 \; 8/3 + 64/5$$

EXAMPLE 3B—

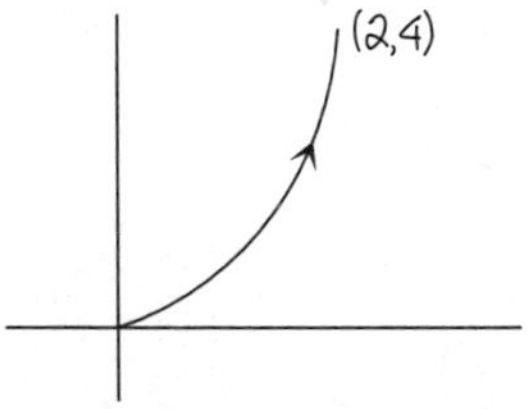

Same **F**.

C_4: $x = 2t$, $y = 4t^2$, $0 \le t \le 1$

$\mathbf{F} = \langle 4t^2, 8t^3 \rangle$ $\mathbf{dr} = \langle 2dt, 8tdt \rangle$

$$\int \mathbf{F} \cdot \mathbf{dr} = \int_0^1 (8t^2 + 64t^4)\, dt = 8t^3/3 + 64t^5/5 \Big[_0^1 = 8/3 + 64/5$$

Hmmmmmmmmmm.

EXAMPLE 3C—

Same **F**.

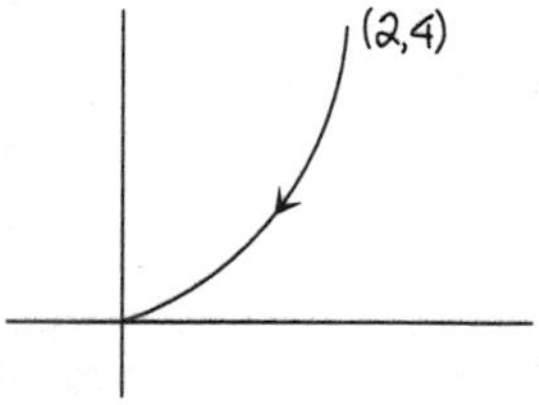

C_5: $x = 2 - t$, $y = (2 - t)^2$, $0 \le t \le 2$

$\mathbf{F} = \langle x^2, xy\rangle = \langle (2-t)^2, (2-t)^3\rangle$, $\mathbf{dr} = \langle -dt, -2(2-t)\,dt\rangle$

$$\int \mathbf{F}\cdot\mathbf{dr} = \int_0^2 -(2-t)^2 - 2(2-t)^4\,dt$$

$$= (2-t)^3/3 + 2(2-t)^5/5\Big[_0^2 = -8/3 - 64/5$$

Hmmmmmmm.

We notice that, if we draw C_3 and C_4, they are exactly the same path. The only difference is the parametrical way we represented the path.

We notice that, if we draw C_5 and C_4 (or C_3), we would get exactly the same path except for the order of points. We would get (0,0) first in one representation and last in the other—the order of points are reversed.

In 3A and 3B, it can be shown that this will always happen. The same is true in 3B and 3C. Let us state the result.

If $C_3 = C_4$, $\int_{C_3} \mathbf{F}\cdot\mathbf{dr} = \int_{C_4} \mathbf{F}\cdot\mathbf{dr}$

If $C_4 = -C_5$, $\int_{C_4} \mathbf{F}\cdot\mathbf{dr} = -\int_{C_5} \mathbf{F}\cdot\mathbf{dr}$

We can extend all of this to 3D.

$\mathbf{F} = M(x,y,z)i + N(x,y,z)j + P(x,y,z)k$

$\mathbf{dr} = dx\,i + dy\,j + dz\,k$

$$\int_C \mathbf{F}\cdot\mathbf{dr} = \int M\,dx + N\,dy + P\,dz$$

EXAMPLE 4—

$\mathbf{F} = \langle 3x^2, xyz, 2z\rangle$. Find $\int_C \mathbf{F}\cdot\mathbf{dr}$ if C is the straight line from (0,0,0) to (2,4,6).

$x = 2t$, $y = 4t$, $z = 6t$, $0 \le t \le 1$. $\mathbf{F} = \langle 12t^2, 48t^3, 12t\rangle$,

$\mathbf{dr} = \langle 2dt, 4dt, 6dt\rangle$.

$$\int_C \mathbf{F}\cdot\mathbf{dr} = \int_0^1 (24t^2 + 192t^3 + 72t)\,dt$$

$$= 8t^3 + 48t^4 + 36t^2\Big[_0^1 = 92$$

We can also find the line integral with respect to the arc s.

$$\int_C \mathbf{F}\cdot\mathbf{ds} = \int_a^b F(x(t),y(t))\sqrt{((x'(t))^2 + (y'(t))^2}\,dt$$

EXAMPLE 5—

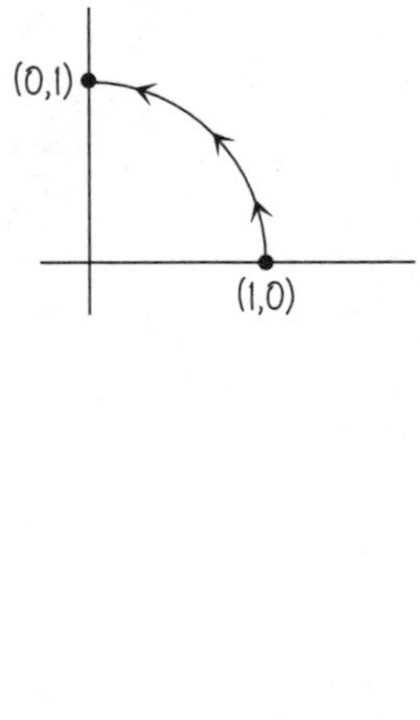

$F(x,y) = 500x^{99}y$, C = from (1,0) to (0,1) on the unit circle (counterclockwise).

We will put the unit circle in parameters the usual way: $x = \cos t$, $y = \sin t$ $\quad 0 \le t \le \pi/2$.

$$\int \mathbf{F}\cdot\mathbf{ds} = \int_0^{\pi/2} 500\cos^{99} t\,\sin t\sqrt{(-\sin t)^2 + (\cos t)^2}\,dt$$

$$= -5\cos^{100} t\Big[_0^{\pi/2} = 5$$

Again, we can extend to three dimensions.

Most of the time, the line integral depends on the path C. However, under certain conditions, the integral depends only on the end points.

In *Calc I,* we studied the fundamental theorem of calculus. Briefly, it stated that, if f(x) is continuous on [a,b] and if F(x) is an antiderivative of f(x), then $\int f(x)\,dx = F(b) - F(a)$.

Let's extend this concept to vector fields. We will just outline the theorems and do some examples. Hopefully, this will allow you to go deeper into this subject.

FUNDAMENTAL THEOREM OF LINE INTEGRALS (F.T.L.I.)

Given

1. (x_1,y_1), an initial point, final point (x_2,y_2), and curve C connecting them in our set. (C is piecewise smooth)
2. $\mathbf{F}(x,y) = \nabla\phi(x,y)$
3. $\mathbf{F} = Mi + Nj$, M, N continuous

Then

$$\int_C \mathbf{F}(x,y) \cdot \mathbf{dr} = \phi(x_2,y_2) - \phi(x_1,y_1)$$

Terms

- If this theorem is true, the integral is *independent* of the path.
- If **F** is the gradient of some function $\phi(x,y)$, **F** is said to be *conservative*.
- ϕ is called the *scalar potential function* for **F**.

EXAMPLE 6—

$\mathbf{F} = 2xy^3i + 3x^2y^2j$. Curve C begins at (3,1) and ends at (1,4).

Find $\int_C \mathbf{F} \cdot \mathbf{dr}$.

A little later, we will actually find a way to show that ϕ exists. You can show that $\phi = x^2y^3$ (show grad $\phi = \mathbf{F}$). By the F.T.L.I.,

$$\int_C \mathbf{F} \cdot \mathbf{dr} = \Phi(1,4) - \Phi(3,1) = 64 - 9 = 55$$

NOTE

This integral can be negative. (Interchange first and last points.)

NOTE ALSO

If we have a closed path, according to the conditions of this fundamental theorem, the integral is 0, since the initial and terminal points would be the same. OK, OK, cut out the baloney. When do we have a conservative **F**, and how do we get it?

THEOREM

$\mathbf{F}(x,y) = M(x,y)i + N(x,y)j$

F conservative if and only if $M_y = N_x$.

To summarize, $\int_C M\,dx + N\,dy$ is independent if and only if $M_y = N_x$.

NOTE

We may never be able to get ϕ, because we just can't find the antiderivative of some things.

EXAMPLE 7—

Let $\mathbf{F} = 2xy^3i + 3x^2y^2j$. Let's find ϕ, if possible.

$M = 2xy^3 \quad N = 3x^2y^2$

$M_y = 6xy^2 \quad N_x = 6xy^2$

Aha! ϕ exists!!!! Can we find it? Yes, in this case.

$$\text{grad}\ \phi = \phi_x i + \phi_y j$$
$$= Mi + Nj$$

$\phi_x = 2xy^3$

So

$\phi = x^2y^3 + f(y)$

Like in a double integral, integrate with respect to x. You hold y constant. The constant of integration could have y's in it, since you are integrating with respect to x only.

$\phi'_y = 3x^2y^2 + f'(y) = N = 3x^2y^2$

So $f'(y) = 0$. So $\phi = x^2y^3$. (For convenience, we can let the numerical constant = 0, since it will cancel out in evaluation.)

EXAMPLE 8—

Let's try one more slightly more complicated example.

Let $\mathbf{F} = (2xy^2 + 3x^2)\mathbf{i} + (2x^2y + 4y^3)\mathbf{j}$. $M = 2xy^2 + 3x^2$, $M_y = 4xy$. $N = 2x^2y + 4y^3$, $N_x = 4xy$. $M_y = N_x$. ϕ lives!!!!

$N = \phi_y = 2x^2y + 4y^3$. $\phi = x^2y^2 + y^4 + g(x)$.

$\phi_x = 2xy^2 + g'(x) = M = 2xy^2 + 3x^2$.

Soooooooooooooo, $g'(x) = 3x^2$. $g(x) = x^3$. Therefore, $\phi = x^2y^2 + y^4 + x^3$.

These problems can be made harder by making the antiderivatives harder, but if you study the easier ones, you'll get the idea.

NOTE

We can extend this also to three variables in 3D.

Let's do one more longish, three-part example.

EXAMPLE 9A—

Let $\mathbf{F} = \langle 2xy^3, 3x^2y^2 \rangle$. C: $x = t$, $y = t^2$; $0 \le t \le 2$.

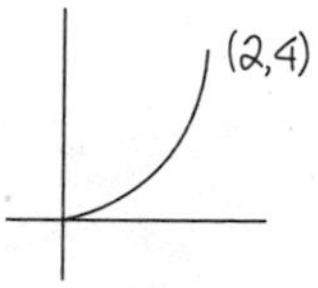

$$\int \mathbf{F} \cdot d\mathbf{r} = \int \langle 2t^7, 3t^6 \rangle \cdot \langle dt, 2t\,dt \rangle = \int_0^2 8t^7\,dt = t^8 \Big|_0^2 = 256$$

EXAMPLE 9B—

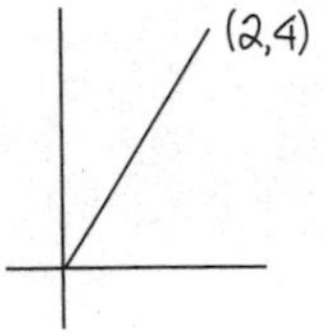

Same **F**, C: x = t, y = 2t; $0 \le t \le 2$.

$$\int \mathbf{F} \cdot d\mathbf{r} = \int \langle 16t^4, 12t^4 \rangle \cdot \langle dt, 2dt \rangle = \int_0^2 40t^4\, dt = 8t^5 \Big[_0^2 = 256$$

Both are 256.0 ps. I forget to test to find if there is a ϕ.

EXAMPLE 9C—

This is the example we did before!!!!! ϕ is x^2y^3!!!!!! So

$$\int \mathbf{F} \cdot d\mathbf{r} = \phi(2,4) - \phi(0,0) = 2^2 4^3 - 0 = 256$$

We can save a lot of time by not having to write in parametric form. All we need is to find ϕ and evaluate. (Once there is a ϕ, it is independent of path and depends only on end points.)

CHAPTER 10

GREEN'S THEOREM

We will now start a series of remarkable theorems that will be presented in much greater detail in later courses.

The first is *Green's theorem.* We will state the theorem, explain it, and give several examples.

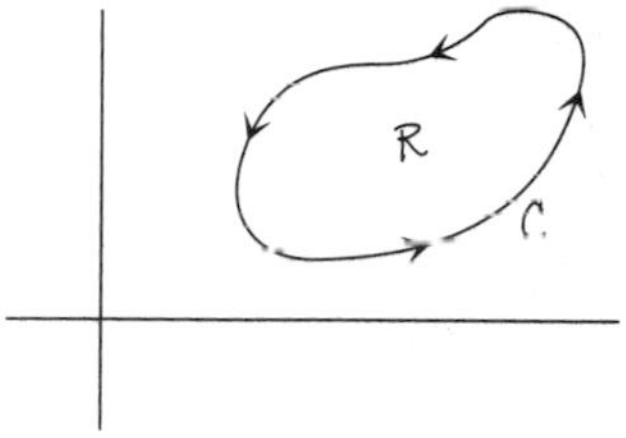

Given vector field F = M(x,y)i + N(x,y)j, M,N have continuous first partial derivatives in an appropriate region containing R (and C). Let C be a closed, simply connected, piecewise smooooth curve in a plane. C is the boundary of R, also in the plane.

Then

$$\oint_C M(x,y)\,dx + N(x,y)\,dy = \iint (N_x - M_y)\,dA$$

1. This is an amazing theorem. To find the value of this integral, you either look at it on its boundary (not peeking at the inside at all) or look on the inside. Both values will be the same!!!!!!! Unbelievable!!!!!!!!!

2. If F(x,y) is conservative, we know that $M_y = N_x$. By the right-hand side, $N_x - M_y = 0$. This means

Green's theorem is used for *nonconservative* fields.

3. We will now do three examples: one showing both parts are equal, one with one side, and one with the other. In general, the right side is shorter, since it is one double integral. The boundary can be many pieces. Sometimes one side or the other is not integrable, in which case you use the one that is, of course.

EXAMPLE 1—

We will verify both parts of Green's theorem for $\mathbf{F} = \langle x^2y, x^3 \rangle$ for the following region.

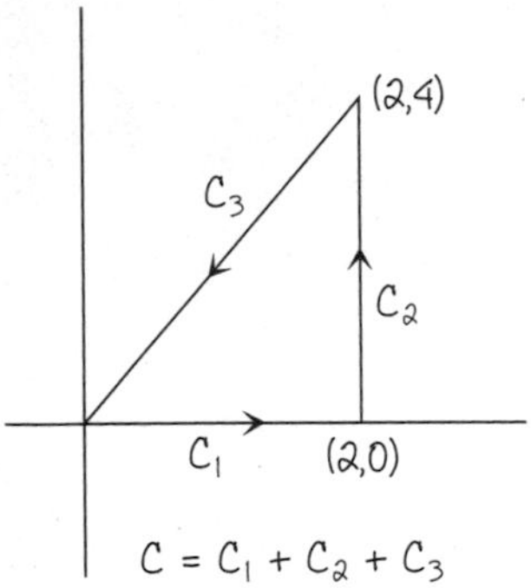

$M = x^2y \quad N = x^3$

On C_1, $x = t$, $dx = dt$; $y = 0$, $dy = 0$.

$M = 0 \quad N = t^3$

but $N\,dy = 0$, so the integral on $C_1 = 0$.

On C_2, $x = 2$, $dx = 0$; $y = t$, $dy = dt$, $0 \le t \le 4$.

$M = 4t \quad N = 8$

$$\int_{C_2} M\,dx + N\,dy = \int_0^4 (8(0) + 8)\,dt = 32.$$

On C_3, $x = 2 - t$, $dx = -dt$; $y = 4 - 2t$, $dy = -2dt$.

$M = 2(2 - t)^3 \quad N = (2 - t)^3$

$$\int_{C_3} \langle 2(2-t)^3, (2-t)^3 \rangle \cdot \langle -dt, -2dt \rangle = \int_0^2 -4(2-t)^3\,dt$$

$$= (2-t)^4 \Big[_0^2 = -16$$

Therefore, if $C = C_1 + C_2 + C_3$,

$$\oint M\,dx + N\,dy = 16$$

Let's do the other part of Green's theorem: $M = x^2y$; $M_y = x^2$; $N = x^3$; $N_x = 3x^2$; $N_x - M_y = 2x^2$.

$$\iint (N_x - M_y)\, dA = \int_{x=0}^{2} \int_{y=0}^{2x} 2x^2\, dy\, dx$$

$$= \int_{x=0}^{2} 2x^2y \Big[_{y=0}^{y=2x}\, dx$$

$$= \int_{x=0}^{2} 4x^3\, dx$$

$$= x^4 \Big[_0^2 = 16$$

So Green's theorem is verified. In this case, both sides are about equal. The line integrals are a little longer, but double integrals are probably a little harder. Let's do two more problems, each one where one side is easier to evaluate.

EXAMPLE 2—

$$\oint_C (x^8 - y^2)\, dx + (2x - 5y)\, dy$$

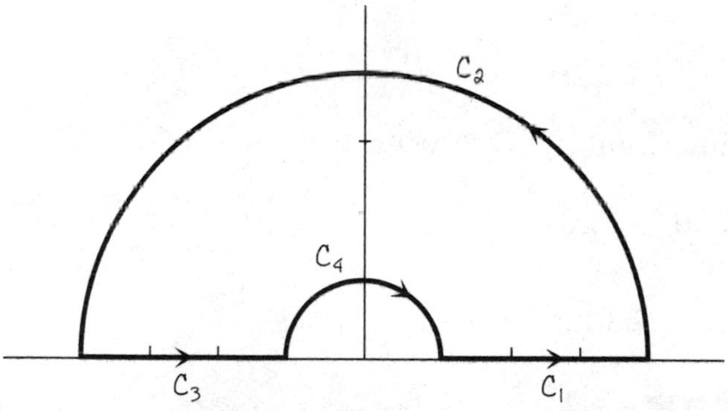

$C = C_1 + C_2 + C_3 + C_4$

To do the line integral on each of the four regions, using parameters on each would be very messy. However, let's use the right half of Green's theorem. We will also use polar coordinates.

$M = x^8 - y^2 \qquad M_y = -2y \qquad N = 2x - 5y \qquad N_x = 2$

$N_x - M_y = 2 + 2y \qquad 1 \le r \le 4$

$\quad = 2 + 2\,r \sin\theta \qquad 0 \le \theta \le \pi$

$$\iint (N_x - M_y)\, dA = \int_{\theta=0}^{\pi} \int_{r=1}^{4} (2 + 2r \sin\theta) r\, dr\, d\theta$$

$$= \int_{\theta=0}^{\pi} r^2 + \frac{2r^3}{3} \sin\theta \Big[_{r=1}^{4} d\theta$$

$$= \int_{\theta=0}^{\pi} (15 + 42 \sin\theta)\, d\theta$$

$$= 15\theta - 42 \cos\theta \Big[_{0}^{\pi} = 15\pi + 84$$

Let us examine $\oint_C x\,dy$. By Green's theorem, $M = 0$, $N = x$, and $N_x = 1$.

$$\oint M\,dx + N\,dy = \iint_R (N_x - M_y)\, dA = \iint 1 dA = A$$

We get the area. Similarly,

$$\oint -y\,dx = A$$

Adding and dividing by 2, we get

$$A = \frac{1}{2} \oint x\,dy - y\,dx$$

Any can be used for the area!!

EXAMPLE 3—

Let's find the area of the ellipse $x^2/a^2 + y^2/b^2 = 1$. Most single-integral or double-integral ways of finding the area of the ellipse are messy and long.

However.....we will use $\frac{1}{2}\oint x\,dy - y\,dx$.

An ellipse in parameters is $x = a \cos t$, $dx = -a \sin t\,dt$, $0 \le t \le 2\pi$.
$y = b \sin t$, $dy = b \cos t\,dt$.

$$\frac{1}{2}\oint_C x\,dy - y\,dx$$

$$= \frac{1}{2}\int_0^{2\pi} a \cos t(b \cos t\,dt) - b \sin t(-a \sin t\,dt)$$

$$= \tfrac{1}{2}ab \int_0^{2\pi} dt = \pi ab$$

There are two other interesting applications we will talk briefly about.

1. Let **T** = unit tangent vector to a curve C.

T = (dx/ds)i + (dy/ds)j

Then

$$\oint_C \mathbf{F} \cdot \mathbf{T}\,ds = \iint_R N_x - M_y\,dA$$

This measures the circulation (or flow) of F around C.

2. Let **N** = unit normal to C.

N = (dy/ds)i – (dx/ds)j

Then

$$\oint \mathbf{F} \cdot \mathbf{N}\,ds = \iint_R M_x + N_y\,dA$$

This is called the flux integral and measures the amount a substance, say a fluid or gas, enters or leaves a region surrounded by C. Positive flux means the substance is leaving. Negative flux means the substance is entering (or more of it is entering than leaving).

CHAPTER 11

OTHER THEOREMS: STOKE'S, DIVERGENCE, ETC.

We will now list, talk about, and do several examples of the remaining topics.

1. Del operator

   $$\nabla = \frac{\partial}{\partial x}\,i + \frac{\partial}{\partial y}\,j + \frac{\partial}{\partial z}\,k$$

2. The curl, a vector of the vector field $\mathbf{F} = Mi + Nj + P(x,y,z)k$

   $$\textbf{curl F} = \nabla \times F = (P_y - N_z)i + (M_z - P_x)j + (N_x - M_y)k$$

   or

   $$\begin{vmatrix} i & j & k \\ \partial/\partial x & \partial/\partial y & \partial/\partial x \\ M & N & P \end{vmatrix}$$

3. Stoke's theorem

   $$\oint_C \mathbf{F} \cdot \mathbf{dr} = \iint_S (\textbf{curl F}) \cdot \mathbf{N}\, dS$$

   This is another amazing theorem. We can find the value of the integral by examining the edge of a surface or the surface itself!!!!

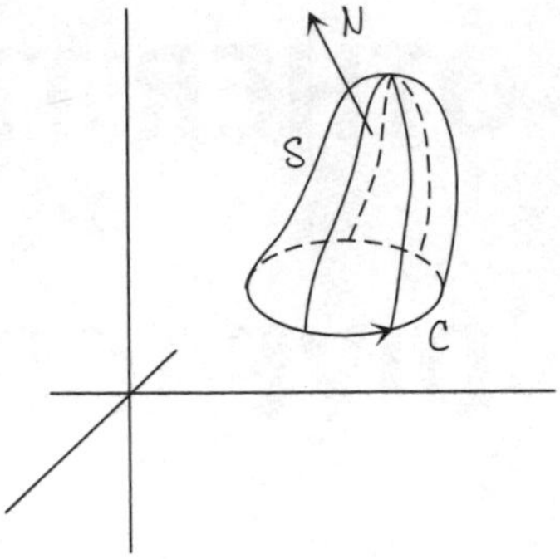

4. Divergence (scalar) of **F**

   div **F** = $M_x + N_y + P_z$

5. Gauss's divergence theorem for surface S for 3D region R

   $$\iint_S \mathbf{F} \cdot \mathbf{N}\, dS = \iiint_R \text{div}\, \mathbf{F}\, dV$$

   Again, another amazing result. We can find the value of the integral by examining the surface, without knowing what's inside, or using the inside, and the results will be the same!!!!! Amazing!!!! It's sort of a 3D Green's theorem.

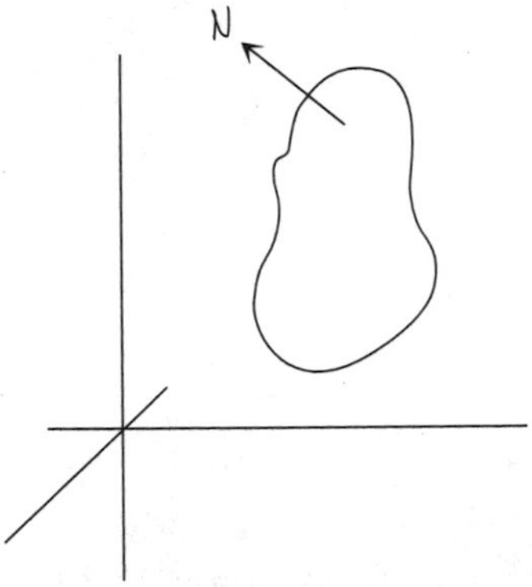

Let us do a few examples.

EXAMPLE 1—

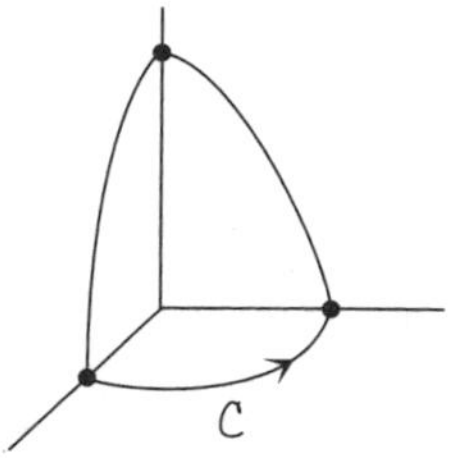

Let us verify Stoke's theorem for the surface $z = 16 - x^2 - y^2$. For C, trace in x-y plane ($z = 0$) $x^2 + y^2 = 16$. $F = 3yi + 4zk + 6xk$.

$$n = z_x i + z_y j + k/\sqrt{(z_x^2 + z_y^2 + 1)}$$

$$= 2xi + 2yj + k/\sqrt{(4x^2 + 4y^2 + 1)}$$

$$\textbf{curl F} = \begin{vmatrix} i & j & k \\ \partial/\partial x & \partial/\partial y & \partial/\partial z \\ 3y & 4z & 6x \end{vmatrix} = -4i - 6j - 3k$$

M = −4, N = −6, P = −3

$$\iint \textbf{curl F} \cdot \textbf{n}\, dS = \iint -Mz_x - Nz_y + P\, dA$$

$$= \iint -(-4)2x - (-6)2y - 3\, dx\, dy$$

$$= \iint 8x + 12y - 3\, dx\, dy$$

$$= \int_{\theta=0}^{2\pi} \int_{r=0}^{4} (8r\cos\theta + 12r\sin\theta - 3)r\, dr\, d\theta$$

$$= \int_0^{2\pi} 4r^2\cos\theta + 6r^2\sin\theta - 3r^2/2 \Big[_{r=0}^{4}\, d\theta$$

$$= \int_0^{2\pi} (64\cos\theta + 96\sin\theta - 24)\, d\theta$$

$$= 64\sin\theta - 96\cos\theta - 24\theta \Big[_0^{2\pi}$$

$$= -48\pi$$

In the x-y plane, **F** reduces to $3yi$.

$$\oint \textbf{F} \cdot \textbf{dr} = \oint_C 3y\, dx$$

$x^2 + y^2 = 16$. For C:
$x = 4\cos t$,
$dx = -4\sin\theta\, d\theta$;
$y = 4\sin t, dy = 4\cos t\, dt$.

$$= \int_{t=0}^{2\pi} 3(4\sin t)(-4\sin t\, dt)$$

$$= \int_0^{2\pi} -48 \sin^2 t \, dt$$

$$= \int_0^{2\pi} -24(1 - \cos 2t) \, dt = -48\pi$$

The two answers agree, as they must.

EXAMPLE 2—

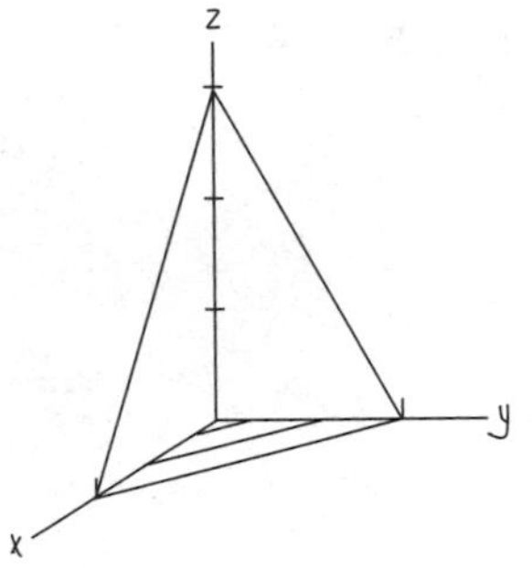

Let us verify the divergence theorem for $\mathbf{F} = x^2 i - 2xy j + 10zk$ for the region bounded by $2x + 2y + z = 2$ and xy,xz,yz planes.

PART A

$M = x^2$, $N = -2xy$, $P = 10z$.

$\nabla F = M_x + N_y + P_z = 2x - 2x + 10 = 10$

$$\iiint_V 10 \, dV$$

Earlier we showed the volume $V = (1/6)abc = (1/6)(1)(1)(2) = 1/3$. The triple integral $= 10(1/3) = 10/3$.

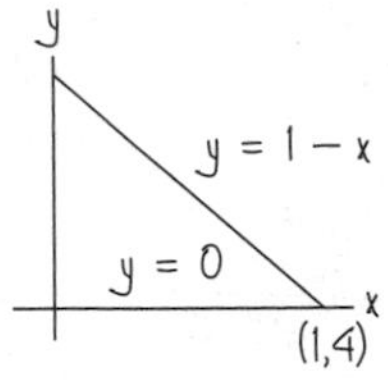

PART B

Surface	F	n	F · n
yz plane – $x = 0$	$10k$	$-i$	0
xz plane – $y = 0$	$x^2i + 10k$	$-j$	0
xy plane – $z = 0$	$x^2i - 2xyj$	$-k$	0
$2x + 2y + z = 2$	$x^2i - 2xyj + 10zk$	$\dfrac{z_x i + z_y j + k}{\sqrt{(z_x^2 + z_y^2 + 1)}}$	$\dfrac{2i + 2j + k}{3}$

$M = x^2$ and $N = -2xy$ and $P = 10z = 10(2 - 2x - 2y)$

$$\iint \mathbf{F} \cdot \mathbf{n}\, dS = \iint (-Mz_x - Nz_y + P)\, dA$$

$$= \int_{x=0}^{1} \int_{y=0}^{1-x} \left(-x^2 \frac{2}{3} + \frac{4xy}{3} + 10(2 - 2x - 2y)\right) dy\, dx$$

$$= \int_{x=0}^{1} -\frac{2}{3} x^2 y + \frac{2x^2 y}{3} + 20y - 20xy - 10y^2 \Big[_0^{1-x} \, dx$$

$$= \int_{x=0}^{1} 20[1 - x] - 20x[1 - x] - 10(1 - x)^2\, dx$$

$$= -10(1 - x)^2 - 10x^2 + \frac{20x^3}{3} + \frac{10(1 - x)^2}{3} \Big[_0^{1}$$

$$= \frac{20}{3} - 10 + 10 \frac{-10}{3} = \frac{10}{3}$$

NOTE

I know that in the last two examples the integrals were rather easy. The purpose was to give you the idea, not to make the book twice as long.

These theorems are very powerful. They can and should be studied in great depth. But that is for another course and another book.

ACKNOWLEDGMENTS

I have many people to thank.

I would like to thank my wife Marlene, who makes life worth living.

I thank the two most wonderful children in the world, Sheryl and Eric, for being themselves.

I would like to thank my brother Jerry for all his encouragement and for arranging to have my nonprofessional editions printed.

I would like to thank Bernice Rothstein of the City College of New York and Sy Solomon at Middlesex County Community College for allowing my books to be sold in their bookstores and for their kindness and encouragement.

I would like to thank Dr. Robert Urbanski, chairman of the math department at Middlesex, first for his encouragement and second for recommending my books to his students because the students found them valuable.

I thank Bill Summers of the CCNY audiovisual department for his help on this and other endeavors.

Next I would like to thank the backbones of three schools, their secretaries: Hazel Spencer of Miami of Ohio, Libby Alam and Efua Tongé of the City College of New York, and Sharon Nelson of Rutgers.

I would like to thank Marty Levine of Market Source for first presenting my books to McGraw-Hill.

I would like to thank McGraw-Hill, especially John Carleo, John Aliano, David Beckwith, and Pat Koch.

I would like to thank Barbara Gilson, Mary Loebig Giles, and Michelle Matozzo Brucci of McGraw-Hill and Marc Campbell of North Market Street Graphics for improving and beautifying the new editions of this series.

I would also like to thank my parents, Lee and Cele, who saw the beginnings of these books but did not live to see their publication.

I would like to thank three people who helped keep my spirits up when things looked very bleak: a great friend, Gary Pitkofsky; another terrific friend and fellow lecturer, David Schwinger; and my sharer of dreams, my cousin, Keith Ellis, who also did not live to see my books published.

Lastly, I would like to thank my Math 202C, that's my ten o'clock Calculus II, Fall 1990 class at the City College of New York, for finally convincing me to do this book. But let me list them.

John Zabala, Angel Vazquez, Juan Vasquez, Jean Toussaint, Lana Thomas, Juan Suazo, Alton Stewart, Andre Smith, Lindrick Sealy, Herna Sam, John Romero, Edwin Reyes, Roland Philips, Julio Melendez, Jimmy Louis, Marcel Laronde, Vera Kuffour, Yollette Jules, John Hill, Julio Hernandez, Tracia Greenidge, Chantal Germain, Grace Gathungu, Falguni Gandhi, Mark Foster, Dion Edwards, Adler Duverger, Betty Develus, Colin Chin, Jeanmonee Catulle, Yvon Cantave, Frances Boschulte, Angela Boone, Sandra Bataille, Gaber Azzam, Joseph Apollon, Abdelmegid Abdelmegid, and........Veron Lyons.

INDEX